Amir Jafari

Autonomous Wireless Sensor Actuator Network in Logistic Systems

Amir Jafari

Autonomous Wireless Sensor Actuator Network in Logistic Systems

Development and Evaluation

Südwestdeutscher Verlag für Hochschulschriften

Impressum/Imprint (nur für Deutschland/ only for Germany)
Bibliografische Information der Deutschen Nationalbibliothek: Die Deutsche Nationalbibliothek verzeichnet diese Publikation in der Deutschen Nationalbibliografie; detaillierte bibliografische Daten sind im Internet über http://dnb.d-nb.de abrufbar.

Alle in diesem Buch genannten Marken und Produktnamen unterliegen warenzeichen-, marken- oder patentrechtlichem Schutz bzw. sind Warenzeichen oder eingetragene Warenzeichen der jeweiligen Inhaber. Die Wiedergabe von Marken, Produktnamen, Gebrauchsnamen, Handelsnamen, Warenbezeichnungen u.s.w. in diesem Werk berechtigt auch ohne besondere Kennzeichnung nicht zu der Annahme, dass solche Namen im Sinne der Warenzeichen- und Markenschutzgesetzgebung als frei zu betrachten wären und daher von jedermann benutzt werden dürften.

Verlag: Südwestdeutscher Verlag für Hochschulschriften GmbH & Co. KG
Dudweiler Landstr. 99, 66123 Saarbrücken, Deutschland
Telefon +49 681 37 20 271-1, Telefax +49 681 37 20 271-0
Email: info@svh-verlag.de
Zugl.: Bremen, Uni, Diss 2010

Herstellung in Deutschland:
Schaltungsdienst Lange o.H.G., Berlin
Books on Demand GmbH, Norderstedt
Reha GmbH, Saarbrücken
Amazon Distribution GmbH, Leipzig
ISBN: 978-3-8381-1823-9

Imprint (only for USA, GB)
Bibliographic information published by the Deutsche Nationalbibliothek: The Deutsche Nationalbibliothek lists this publication in the Deutsche Nationalbibliografie; detailed bibliographic data are available in the Internet at http://dnb.d-nb.de.

Any brand names and product names mentioned in this book are subject to trademark, brand or patent protection and are trademarks or registered trademarks of their respective holders. The use of brand names, product names, common names, trade names, product descriptions etc. even without a particular marking in this works is in no way to be construed to mean that such names may be regarded as unrestricted in respect of trademark and brand protection legislation and could thus be used by anyone.

Publisher: Südwestdeutscher Verlag für Hochschulschriften GmbH & Co. KG
Dudweiler Landstr. 99, 66123 Saarbrücken, Germany
Phone +49 681 37 20 271-1, Fax +49 681 37 20 271-0
Email: info@svh-verlag.de

Printed in the U.S.A.
Printed in the U.K. by (see last page)
ISBN: 978-3-8381-1823-9

Copyright © 2010 by the author and Südwestdeutscher Verlag für Hochschulschriften GmbH & Co. KG and licensors
All rights reserved. Saarbrücken 2010

Development and evaluation of an autonomous wireless sensor actuator network in logistic systems

Dr.-Ing. Amir M. Jafari

Universität Bremen

Being autonomous is a state of mind, with a centralized mind an autonomous system can degenerate into a central system at any moment.

This work is dedicated to:

Iranians Liberation Movement
and
The brave people who died for their liberty and dignity in this movement

Contents

Abbreviation List i

Preface 1

1. Introduction to the autonomous system 5

   1.1) Central system 6

   1.2) Distributed system 8

      1.2.1) Distributed control system 10
      1.2.2) Decentralized system 11

   1.3) Autonomous system 12

   1.4) Autonomous wireless sensor actuator network 15

   1.5) Logistic system 20

2. Development of AWSAN 24

   2.1) Routing algorithm 25

      2.1.1) Demanded features 25
      2.1.2) Literature review 28
      2.1.3) SCAR development 31

   2.2) Sample number 37

      2.2.1) Sampling theory 37
      2.2.2) Actuator frequency 39
      2.2.3) Discrete control limits and lower limit boundary for the sample number selection 41
      2.2.4) Actuator frequency drift 43
      2.2.5) Actuator frequency drift versus limits interval 45

2.2.6) Sample number selection in CWSAN	....	46
2.2.7) Sample number selection in AWSAN	....	48
2.3) Wireless nodes	....	*51*
2.3.1) Hardware specifications	....	51
2.3.2) Software specification	....	53

3. Simulation and Comparison 55

3.1) Prowler	....	*56*
3.1.1) Introduction	....	57
3.1.2) Modification	....	58
3.2) Energy consumption	....	*60*
3.2.1) Simulation	....	61
3.2.2) Conclusion	....	64
3.2.3) Discussion	....	67
3.3) Robustness comparison by an orchard example	....	*69*
3.3.1) Simulation Preparations	....	70
3.3.2) Conclusion	....	75
3.3.3) Discussion	....	83
3.4) Scalability	....	*84*
3.5) Autonomous or Central	....	*87*
3.5.1) Message number in autonomous and central model	....	90
3.5.2) Comparison by message number	....	91
3.5.3) Compromising based on the total message number	....	92

4. Conclusion 96

References **99**

Publication Lists **103**

Abbreviations List

LAN	Local Area Network
DCS	Distributed Control System
CAN	Control Area Network
SAL	Sensor Actuator Layer
SAN	Sensor Actuator Network
SCADA	Supervisory Control and Data Acquisition
BMS	Building Management System
HVAC	Heating Ventilation and Air Conditioning System
PE	Process Entity
WSAN	Wireless Sensor Actuator Network
AWSAN	Autonomous Wireless Sensor Actuator Network
CWSAN	Central Wireless Sensor Actuator Network
ASAN	Autonomous Wireless Sensor Actuator Network
GPS	Global Positioning System
LPS	Local Positioning System
GPSR	Greedy Perimeter Stateless Routing
RNG	Relative Neighborhood Graph

GG	Gabriel Graph
GEAR	Geographic and Energy Aware Routing
ZRP	Zone Routing Protocol
LCR	Logical Coordinate Routing
SCAR	Sequential Coordinates Routing Algorithm
MLT	Minimum Length Tree
CAV	Channel Attenuation Value
DFS	Deep First Search
RSSI	Received Signal Strength Indicator
PLC	Programmable Logic Controller
ADC	Analog to Digital Convertor
DAC	Digital to Analog Convertor
DMA	Direct Memory Access
DCO	Digitally Controlled Oscillator
ACLK	Auxiliary Clock
MCLK	Main clock
SMCLK	Sub Main Clock
FIFO	First Input First Output
Prowler	Probabilistic Wireless Network Simulator
GUI	Graphical User Interface
EDF	Earliest Deadline First

Preface

This text presents an introduction, development and evaluation of an autonomous system. This system structure is seen as an improved version of a distributed system structure and an alternative to the traditional central structure. Improvements in the communication and digital technology provide conditions which have impacts on the creditability of traditional system organization solutions. The autonomous structure not only combines the capabilities provided by technology to achieve a new way of configuration but also it opens a new horizon with regard to the future of technology. The autonomous or sovereignty concept refers to the constituent entities liberty inside an organization. On one hand balancing between authority and liberty and on the other hand changing the system to limit its higher level of dominance over the entities is an endless path because, apart from its advantages, such desires are rooted in human nature.

This text focuses on self-decision making as a basic constituent and comprehensive element of the autonomous system definition. Following this focal point, horizontal relations between entities replace vertical relations of entities. Interdependency on the information between entities substitutes of entities subordination and dependency on the resources. Because of the removal of the entities dependency on one or few entities inside the system, the reliability of the system increases. Distribution of tasks of the system over entities causes the system effort to become distributed over entities too. The horizontal relation between entities and their independency on each other's resources make the system scalable. In an autonomous system, by moving the decision making level to the peripherals, the system becomes faster in reacting to its environmental changes and more capable of self-adaptation.

This text is organized into three chapters. The outline of the text is as follows. Chapter 1 covers the concept and definition of an autonomous system. It starts by reviewing the central system structure, its properties and limitations. Then it looks at the distributed system as a solution to cope with the central system constrains. The appearance of a distributed system in automation is reviewed in the distributed control system section.

Regarding the development of a distributed structure in political science, in the name of federalism and more precise concept definition in this realm, one eye is kept on federalism as well. The similar terms to the autonomous adjective from different references are introduced and discussed to avoid vagueness and to reach a clear mind about the autonomous concept. In the following, the implementation of an autonomous structure for the sensor actuator network is introduced and based on the feedback model which is offered to the task description of each element. At the end of this chapter, by returning to the general concept of the autonomous system, its application for logistic system is explained. It is discussed how the competition between entities inside the system can turn to a constructive competition over the goals. The outcome of this discussion links the autonomous wireless sensor actuator network with the intelligent container as a required entity for the autonomous logistic system.

Chapter 2 introduces two elements for developing an autonomous structure for a wireless sensor actuator network and an instant node for further practical implementation. The information interdependency of the nodes necessitates direct communication between nodes. A routing algorithm which supports this feature is categorized under target-oriented routing algorithms. In this chapter, a sequential coordinate routing algorithm is introduced and developed for the autonomous network. A method is introduced to find an optimal sample number for wireless sensors inside the autonomous and central network. Wireless communication nature and limited power supply impose conditions on the sampling rate which is considered in these methods. The wireless sensor node introduced at the end of this chapter is one example node to realize this network in practice. This example node specification is used for application simulations in the next chapter.

The third chapter is about the simulation of autonomous and central wireless sensor networks and a comparison of these two structures. Firstly, a simulator and its functionality are introduced. It is stated that for whichever reasons this simulator is chosen, the applied modifications to this simulator are explained. In the first simulation the energy consumption and its distribution in two networks are evaluated and compared. The results of this simulation are formulated with regard to comparative performance and sustainability. The second simulation in this chapter is designed to compare the robustness of these two network structures. Humidity control of an apple orchard is taken as an application for simulation. The dynamic behavior and the sensor values are modeled by state space variable. The noise over the communication channels is modeled by random distributions. Some conclusions are drawn by comparing the system outputs of both networks about robustness and noise effect distribution.

The whole work can be seen with two points of view. From one of the viewpoint this work generally refers to introduction, development and assessment of an autonomous wireless sensor actuator network for an automation system. From another point of view this work is about introduction and evaluation of an autonomous system. These points of view are linked in the way that the autonomous wireless sensor actuator network is a sample and small scale version of autonomous system in general. In fact by this scaling, the macroscopic viewpoint of autonomous system is replaced by the microscopic viewpoint of autonomous sensor network which is easier to analyze. At the end of pervious sections in third chapter, the results are generalized for logistic systems and the standpoints are changed from autonomous wireless sensor actuator network to the autonomous logistic system by some resemblances. At these subsections, the microscopic viewpoint is mapped to macroscopic view.

Following the third chapter regarding the optimal sample number, the scalability and control quality of both networks are assessed. At the end of this chapter the creditability of the results in different conditions is considered. We attempt to answer this question of under which condition(s) one of the network structures has priority over another and achieves advantages. This part of the third chapter returns to the first discussion of information interdependency of the entities. A tradeoff point is introduced by which the decision can be made to choose one of the two possible structures for the network. Chapter 4 summarizes the conclusions of this work.

Acknowledgments

This research was conducted at the Institute for Microsensors, -actuators and -systems (IMSAS). This project was supported financially by the International Graduate School for Dynamics in Logistics at the University of Bremen. I wish to thank my colleagues at IMSAS and IGS. I would like to express my sincere gratitude to my professor, Prof. Dr. Ing. Walter Lang, who provided a good atmosphere to work as a responsible autonomous Ph.D. student and practice this concept alongside developing this system. I want to thank my colleague, Dipl. Ing. Adam Sklorz, and the director of IGS, Dr. Ingrid Rügge, who helped me during my studies to integrate in my new environment. Dr. Tanja Klenk from the Centre for Social Policy Research at the University of Bremen shared her vision and was influential in shaping my understanding about the federalism concept. She reviewed the first chapter as a degree holder in political science. I appreciate her for all her attention to this work. I would like to thank Andrea O'Brien from

Dundalk Institute of Technology who did some of the proof reading of this manuscript. Last, but not least, I wish to express my gratitude to my parents for everything they have done for me.

Amir Jafari
26.01.2010
Bremen, Germany

Chapter 1

Introduction to the autonomous system

Abstract:
To introduce the concept of the autonomous system, this chapter starts by reviewing the central system concept and structure. Then the central system constrains are explained and it is shown how the distributed system is developed to cope with the limitations of the central system. The distributed control system emergence following distributed system concepts is studied. Another term which is used for the distributed control system is decentralized. In one section the usage of this term is explained. The autonomous system concept based on these bases with consideration of its meaning in other realms is discussed. The autonomous structure has already attracted the attention of philosophers in political science in the field of the federal system. In order to obtain a clearer understanding of this term, it is tried to show the difference between autonomous and other common terms. In this survey the self-decision making is offered as a core element of the autonomous concept. Following this concept and considering the point that the autonomous system inherits the distributed system characters, the autonomous wireless sensor actuator network is introduced. In the last section it is explained how the autonomous structure in logistic systems can conduct the competition of the entities from the resources to the goals. In this way the necessity of developing the autonomous objects such as intelligent containers manifested itself.

1.1) Central system

According to Encyclopedia Britannica, a system is "a regularly interacting or interdependent group of items forming a unified whole". Generally three characteristics are counted for a system: structure, behavior and interconnection. Structure is about the composition of items which are called entities hereafter. Behavior refers to the character which the system portrays by itself when it is looked at from outside of the system. Interconnectivity describes the connectivity between the entities inside the system and it is known as a system network. A central system contains an entity which decides about the behavior of the system or plays a switching role in connection between the entities.

In terms of networking there is an entity in the central system whose role is to make connection between entities, e.g. the center receives a message and sends it to the destination. In the central decision-making system, the supreme entity (center) makes decisions for the other entities and for the whole system and has authority over other entities. In political science such a system is also called an authority system [01King]. Decision-making is taken as a factor because it covers many aspects such as control, organization of the entities, task distribution and resource allocations. Referring to the system characteristic, the system network is independent from decision-making structures because in some systems the entities can make decisions for themselves but the data exchange between them is done via a center. Therefore the central system is referred to it is more precise to say from which aspect.

A primary model of the central system is shown in Fig. 1.1. Elazar in [15Elaz] names this model Center-Periphery model and Tanenbaum in [12Tanb] names this model Star. The Center-Periphery model is preferred to Star because it represents the center peripherals' roles which connect to the subordinates. In this figure the center entity is called supreme entity or easily center and the connected entities are subordinated to this center. The arrows show the connection path of the center and entities. The figure shows that the entities are not dependent on or connected to each other directly. In other words, the unification of the system is totally achieved by the center. If the center fails, the system will break apart.

Now suppose that more entities are supposed to attach to the center, since the peripheral number of the center is limited, at one point the center will go out of its peripherals. After this point in order to add more entities to the center one solution is to add more resources and peripherals to the center. This solution is costly because not only does the old center becomes useless but also based on what Tanenbaum in

1. Introduction to Autonomous system

[12Tanb] says, a more powerful center costs more than adding new centers, for example a processor with double capacity costs more than two processors of the same type. Moreover sometimes it is not technically even possible to provide a more powerful center, for example suppose that a system has the most powerful center then a center, with 5 times more power and resources, does not even exist.

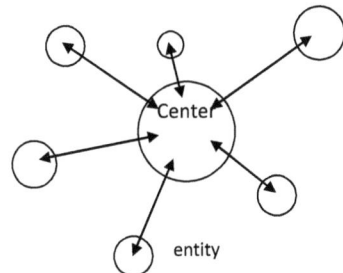

Figure 1.1: Center-Periphery model

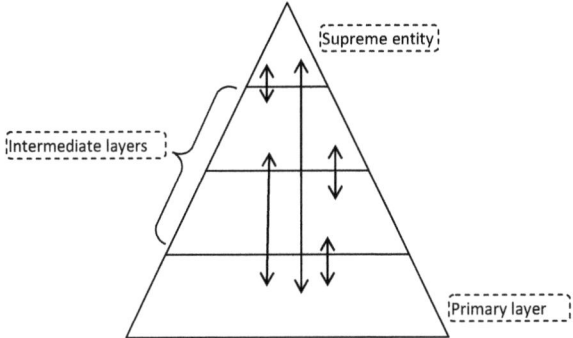

Figure 1.2: hierarchy model

Apart from the dispute whether this solution is practical or not, another solution replaces new centers connected to the main center instead of the entities in Fig. 1.1. Then it is possible to connect the entities to these new sub-centers and make subsystems. In this way the peripheral number and resources increase and a new structure forms which is called hierarchy in [09Duff][03Dres][12Tanb][15Elaz]. This model is illustrated in Fig. 1.2. In this structure the top of the pyramid has the highest authority in decision making and it gradually reduces from top to bottom. As it is shown in this figure by arrows, the connections between the entities and layers are vertical not horizontal. This means that the entities in same layer are independent and unification of each layer is provided by entities in the layer above. In this structure, if the supreme entity on the top of the pyramid fails, the layer below will be disunified and inherently disunification will spread into other layers. Later it will be explained that the problems for scaling the central system mentioned previously have led to develop a Distributed System. An Autonomous system is then introduced, in which unification is achieved by interdependency.

1.2) Distributed system

In the technology domain to overcome the constrains of the central system, particularly scalability [02Wu][03Dres][12Tanb][04Wang], a distributed system is developed. There is no general and common definition for a distributed system [03Dres][12Tanb] for example Tanenbaum in [12Tanb], Wang in [04Wang] and Wu in [02Wu] express different definitions or Zhang in [06Zhan] offers just some features. Wu in [02Wu] says that "distributed" as an adjective refers to distribution of "hardware" such as processor and memory, control and data. On the other hand this system is discriminated from network [02Wu] or parallel system [12Tanb] [03Dres]. Wu in [02Wu] discusses that if the entities do not have close coordination with each other they are just networked and they do not form system, whether distributed or not, e.g. Local Area Network (LAN). A Parallel System differs from a distributed system since in a parallel system all the subsystems are subjected to solving a single task [12Tanb]. Dressler in [03Dres] follows the same definition as Tanenbaum in [12Tanb]. The definition (1) by Tanenbaum is just about the computer distributed system but another definition element can be substituted to generalize the definition, for example "computers" can be replaced by "subsystems" or "system" may be used instead of "computer".

Definition (1):

"A distributed system is a collection of independent computers that appear to the users of the system as a single computer."

This definition covers two aspects, internally and externally. The "independent computer", instead of entity, expresses that hardware is distributed inside of the system. In other words, none of the entities is dependent on the resources of other entities. The external aspect offered with this definition is that the users see the system service or behavior as a "single system" [12Tanb] or centralized controlled system [03Dres] from outside of the system.

Some features and properties are pointed out in different references for the distributed systems. If a system includes such features it could be said that it is distributed [02Wu][12Tanb][03Dres].

1) **Hardware distribution:** The subsystems or entities have their own resources e.g. processor and memory. In other words it sounds as if the center resources are divided into parts and each entity takes one of the parts.
2) **Reliability:** The system should not be totally dependent on one subsystem or entity functionality. This implies that if one of the subsystems fails, the rest of the system continues to function.
3) **Scalability:** The system has possibility to scale in three terms: size which is defined by the number of subsystems and entities, resources geographical distances and administratively scale.
4) **Transparency:** From external users or applications point of view, the collection of subsystems and entities must look like a coherent single system in order to be called a system. The differences between subsystems or entities with their communication are hidden. For example for working with internet, there are different servers and routers and so on but on the internet browser we see all of them as one coherent system.
5) **Connectivity:** Connecting users or applications with the resources. This feature consists of more than what was already described about the interconnection between entities in system definition. Regardless of entities or subsystems interconnection, a distributed system provides the connectivity between applications or users with the resources.

Decision-making or similarly controlling a structure in a distributed system is controversial. Dressler in [03Dres] claims that a specific control process is dynamically allocated to one of the entities in a distributed system. On the contrary Wu in [02Wu] says the "control" is distributed as well. Zhang in [06Zhan] categorizes the distributed system into three groups: "open", "closed" and "in-between". A completely open systems do not have a controller e.g. internet, and in a closed systems an entity has complete control over the entire system. An in-between system is neither closed nor open. In other words the system behavior is not completely without any supervisory guidance and it is not totally curbed by an entity. These categories are conceptually the same as King's categories in [01King] for a federal system: centralist, decentralist and balance federalism. Centralists seek unification around a center and give priority to the central decision-making just like the closed systems offered by Zhang. Decentralists are based on anarchism seek the maximum liberty and abolish the central control over the system. The balance federalists, like the "in-between" system, seek balance between these two systems.

1.2.1) Distributed control system

In industrial automation, the distributed system emerges as a Distributed Control System (DCS). A bunch of sensors or actuators of a process can be bundled by a controller. The sensor and actuator connects to controller in Star or Bus topology [06Zhan] with Profibus, Fieldbus, Control Area Network (CAN), LonWork protocol, etc [05Milan][06Zhan][07Lev][08CIS]. In future Sensors and Actuators will be referred as Sensor Actuator Layer (SAL) and their network is called Sensor Actuator Network (SAN). Zhang in [06Zhan] claims that DCS does not usually include supervision control and it should be designed so that it performs its job without human intervention in a normal situation. But when a DCS with Supervisory Control and Data Acquisition (SCADA) are combined then an operator can intervene in the decision making process. Figure 1.3 shows an example of a SCADA system. In this figure SAN1 represents bus topology and SAN2 represents the star topology. The controllers which are networked make a layer, called a "controller layer". Above the controller layer there is a supervisory, monitoring and archiving layer in the SCADA system.

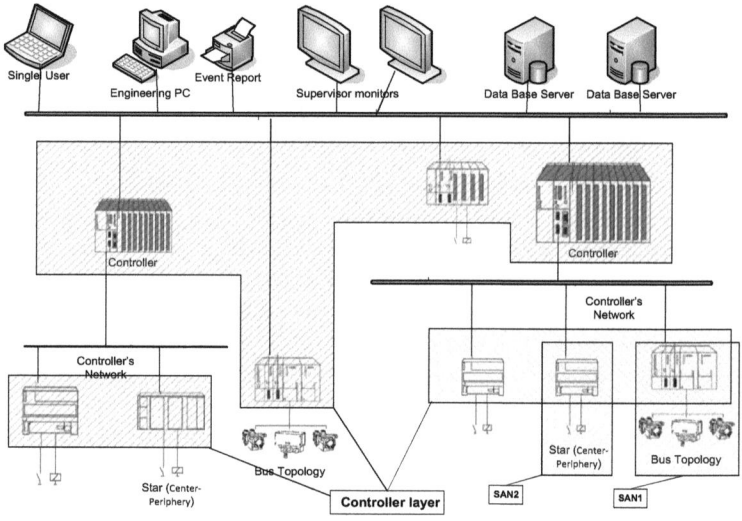

Figure 1.3: SCADA schematic

Internal structure of DCS in terms of interconnections is hierarchical [05Milan] but one of its goals is to be heterogeneous to incorporate different types and trademarks of sensors, actuator and controllers [05Milan][06Zhan]. One of the main properties of distributed system is to maintain diversity (in political science it refers to pluralism [01King][15Elaz]). To achieve this goal the sensor, actuators and controllers with different trademarks should support the same communication protocol in their layer e.g. Profibus, Fieldbus, CAN or Ethernet in the controller layer. On the other hand, in this way it maintains the option to scale the system. Some characteristics the same as mentioned for distributed system are stated for DCS in different references such as [05Milan][06Zhan][07Lev]:

1. Distribution of hardware and function
2. Transparency in order to look like a single system
3. Scalability in terms of subsystems number, geographical and administrative distribution.

With regard to decision-making or correspondently controlling the system, DCS internal structure is almost hierarchical, although the aforementioned ambiguity of a distributed system is inherited in the controller layer as well. In the sensor-actuator layer the sensors send their data to the controller and it processes the data, makes decisions and sends an order to the actuators, which as introduced before is called Center-Periphery model. The combination of Center-Periphery subsystems establishes hierarchical structure (Fig. 1.3). In the controller layer if a decision is dependent on the information of other controllers or sensors and actuators in other groups, it is made by a controller in the upper layer (hierarchical) or the information could be fetched by the controller in the same layer to make decision (nonhierarchical).

DCS is used in a Building Management Systems (BMS) incorporating a Heating Ventilation and Air Conditioning (HVAC) system with the same structure and characteristic [08CIS]. In some literatures and references e.g. [07Lev] [09Duff] such a system is called a decentralized control system and the same characteristics are associated with it, therefore in this work these two systems, distributed and such decentralized control system, are taken to be the same. Decentralized systems are discussed briefly from another point of view in below in order to explain about the usage of the "decentralized" adjective in relation to a system.

1.2.2) Decentralized system

King in [01King] introduces a Decentralist federalism doctrine which is based on the anarchism concept and he looks at it as an approach for federalism. These federalists are aimed towards abolishing any kind of center in the system. In contrast to King, Elazar in [15Elaz] believes that decentralized is not the proper name for such a system because it presupposes that there is a center which makes the system decentralized. From another point of view, if we look at configuring the system structure from the bottom up, decentralized does not mean anything because there is no meaning for decentralized entity as basic constructive element of the system. But if a system is considered from top to bottom and it had already been configured by a center, decentralized implies dividing the center and distributing it over the subsystems. In fact the difference between King and Elazar is the difference between the points they are standing at and from where they are looking at the system. Elazar in [15Elaz] considers the system from the bottom to top and King in [01King] looks at the system from top to bottom. This difference is clearly seen when Böse in [47Böse] mentiones:

> "Therefore, decentralization of the decision making process from the total system to the individual system elements is a specific criterion of autonomous control."

From above sentence it can be seen that direction of decentralization movement is from "total system" which represents top layer in central system to the "individual system elements" which indicates the bottom.

Considering these two points of view is important because they lead to make different definitions and constrain realization. Suppose that on one occasion a house is built from the ground up and on another time the house is renovated and restructured; in each case the conditions and problems are different. Restructuring the system presents limitations which cannot even exist when a system is built from base. On another hand, decentralization process for example mentioned by Böse in [47Böse] does not even exist when it is looked from bottom to top. In this work the system structure is seen from the bottom up, therefore the "decentralize" adjective is not applied. This viewpoint is the same as King's point of view in [01King] when he categorized federalism into three groups, namely centralist, decentralist and balance federalism. He sees the centralist federal as unification in diversity and like American federalism it is made of a sovereign entity from bottom to top.

1.3) Autonomous system

Following the distributed control system which includes ambiguity about its internal structure, the autonomous system concept is considered as complementary. But before developing the concept of the autonomous system, we look at the literatures to see what is inferred with "autonomous".

1. Introduction to Autonomous system

In the domain of political science federalism discusses how the political organization of the society deals with sovereignty which is synonym to autonomy in this realm. Based on King [01King] and Elazar [15Elaz] arguments the autonomous adjective returns to who makes the decision. In the federal system, the decision making tasks is distributed between the federal government and locals. In addition to theoretical discussions, there are some historical events which support such interpretation about the autonomous concepts. As an example, the dispute over the "Import Tax" in the nineteenth century in the United States can be examined. The federal government made a decision to put a tax on imported goods to support its own local products but in the south of the United States the states disagreed because their economy was based on agriculture not industry. They claimed that the federal government does not have the right to make such a law over the states' heads and the states have the right to deny.

As another example, the slavery issue in the middle of the nineteenth century can be named. The southern states believed that slavery was part of their life-style and the federal government did not have the right to abandon or confine it. This was the one of disputes over decision-making rights. The next level dispute over decision-making rights was the secession. When Lincoln won the presidential election, some of the southern states declared secession and established the "Confederation of America". Lincoln claimed that they were united and the states did not have the right to make the decision for secession by themselves. The seceded states called their new collaboration a confederation to represent the right of secession by states. In fact the right to make decisions about secession from system by entities is the difference between freedom in liberalism and anarchism [01King].

Independency

Back to the technology domain, "Autonomous Process Entity (PE)" in [02Wu] is referred to as an entity which has separate physical memory and significant message transmission delay time. It can be seen that Wu in [02Wu] by this expression refers to an autonomous entity as an entity which is independent in resources from other entities. Independency in resources is one of the pillars to self-decision making and self-controlling because if an entity becomes dependent on something outside of itself, it could be controlled through that dependency as well, which contradicts with self-decision making and self controlling. In some control theory texts the system which is only dependent on its own states and does not have input and output is called autonomous system e.g. [10Dull]. Although such an interpretation of the "autonomous" adjective is related to self-decision making in a weak sense, but it is not comprehensive.

Automation

Another controversial usage of the autonomous concept is seen by Wang in [04wang]. Since this kind of interpretation about autonomous property is seen in other works, this concept is discussed here to avoid of ambiguity. He claims:

> "The sensor data and actuator feedback data is directly fed to the controller over the network. This can also be called the autonomous networked control system. The supervisory controller is not a human in this case."

From this it can be inferred that he sees the autonomous system as equal to a system in which human has no intervention. Simply if the Oxford dictionary is taken as reference for the meaning of the words, the expressed properties for an autonomous system lie under the "Automatic" adjective not "Autonomous". When a system is automated, it would fulfill its goal without human intervention. Now the question is whether these two definitions are equivalent or not?

Since the above expression is stated in the technology domain, firstly we will discuss it with regard to the technology domain. Logically, automatic is equivalent to the autonomous when one of these is necessary and sufficient for another one. Suppose that a system (or say a machine) is autonomous (like a robot), it implies that the system makes decisions including the controlling decisions for itself; then it will be free from an external entity involvement including human. It means that being autonomous is sufficient for being automated. In this case if we extend the domain of application for autonomous concepts and apply it to a human organization, these sufficient conditions become invalid as well. As proof, consider a department in which its employees work as autonomous entities and establish an autonomous system, in this case it does not make sense to say they are automated.

Being autonomous is not a necessary condition for being automated. A system can be automated (without human involvement) but without being autonomous in the way that it gets its decisions from outside of itself. For example in Fig. 1.3 in each SAN, the controller makes decision and sends the decisions to the actuators (Center-Periphery model).

Sufficiency of being autonomous for being automated in the technology domain is implicitly pointed out by Dressler in [03Dres] with this term: *"The higher the system's autonomy, the lower the human involvement"*. It means that from autonomous system it is possible to reach the point that human involvement is ignored, but as it is explained it is not bilateral. Perhaps such an expected outcome of an autonomous system is the reason for taking the automated system as an autonomous one. To avoid any ambiguity and clarification for the applied adjective we use "automated" for the methods and concepts referring to using machines instead of human and "autonomous" for the concept which refers to the self-decision making.

Authority versus Autonomy

Duffie in [09Duff] introduces Hierarchy, Heterarchy, Responsible Autonomy and Anarchy as system structures. In these categories, differences between entities "authority" become lower from Hierarchy to Anarchy. In Hierarchy a supreme entity has authority over the others and it makes the decision. In Heterarchy and Responsible Autonomy the entities are not subordinate of each other and they make their own decision. Like Elazar [15Elaz] matrix model for federal system, in a responsible autonomy the entities are limited to some boundaries. On other hand they are accountable for their outcome. Using "authority" and "subordination" by Duffie to compare different systems implies that autonomy generally and conceptually refers to self-decision making.

Dressler in [03Dres] introduces the self-organized system concepts. He claims that for establishing a self-organized system the entities inside the system should be autonomous and the structure and functionality of the system "emerges" from interactions of these autonomous components. He looks at the system components from external point of view and he does not concern with internal structure of the entities. He does not deeply and conceptually discuss about autonomous property but in another discussion in the same reference he addresses:

> "Sensors gather information about physical world, while actuators make decisions and then perform appropriate action upon the environment, which allows a user to effectively sense and act from distance."

Regardless of the mentioned advantage (Connectivity of users and resources) which is inherited from distributed systems the concept of self-decision making from autonomous concepts is inferred here from Dressler's viewpoint. The actuators make decisions by themselves on what to do.

Another reference which enhances the perception of self-decision making from the "autonomous" adjective is [47Böse]. In this reference, the autonomously controlled logistic systems are defined with the below expression:

> "Autonomously Control in logistic systems is characterized by the ability of logistic objects to process information, to render and to execute decisions on their own."

In this expression it can be seen that firstly the characterization refers to the system constituent entities. Secondly the capability of processing information by entities can be interpreted as independency in resources. Thirdly execution of decision on their own is the same property of not being subordinate of other entities and prioritize the decisions. Regarding to these concepts, in this work the self-decision making is taken as essential property of autonomy. As it is already mentioned in distributed system properties, placement of decision making in level of the constituent entities makes the system to response the surrounding environmental changes faster, more robust and reliable. These properties are evaluated in next chapters.

1.4) Autonomous wireless sensor actuator network

Pursuing Distributed systems and DCS, we consider the SAN in Fig. 1.3 in order to develop a distributed SAN. As has been already mentioned SAN has a center-periphery (Star) structure. The motivation for moving from the central to distributed SAN is the same as the aforementioned advantages of distributed systems i.e. reliability, scalability and connectivity. To be more precise about the structure of SAN, the autonomous adjective is attached to the SAN. As it is explained by now, an Autonomous entity is defined as an entity which makes decisions for itself and an autonomous network is the composition of autonomous entities. In the case of entities in a network forming a unified whole and have a close cooperation to achieve a goal, it is called autonomous system.

In a central network the nodes are dependent on the center by two terms: firstly the decision-making and secondly exchanging the data, therefore if the center fails, the system fails totally. Based on the reliability property of the distributed system when a node fails, the rest of the system should continue to operate and the system should not fail completely. By giving the right to each entity to make decisions for itself one of the dependency channels to the center is cut off and the reliability of the system enhances. The autonomous entity for making decisions may need the data and information from the other entities. In this sense they are dependent on each other and interdependency emerges. The resources independency and interdependency is implicitly expressed by Böse in [47Böse]. Through this "interdependency" they can be controlled or have influence on each other, therefore it should be considered that each entity makes its data accessible for others in this system. This exchanging data between the entities is referred to as Cooperation in an autonomous system.

1. Introduction to Autonomous system

The Autonomous entity can establish the interconnections with two architects: Mesh or Star topology. In the star topology there is an entity through which all data exchange is done, like Fig. 1.1. This fact does not interfere with the self-decision making of the entities directly but the entities become dependent on the special entity. This dependency of interconnections weakens the reliability of the system, because of this if the central entity fails all other entities will lose contact with each other and the system will fail as well. This consequence opposes the primary goal of developing distributed system for sensor and actuator layer which is to gain distributed system advantages. On the other hand, since the entities become dependent on this particular entity, this entity can take over the control of the system or influence the system behavior through this dependency. The potential of the central entity puts the self-decision making of the system under question. Because of all these consequences and disadvantages, the mesh topology remains the only option for interconnection in autonomous sensor actuator network.

Before emerging wireless communication, implementing mesh topology for SAL was complicated and costly, because all the sensors and actuators should be connected to each other while in Center-Periphery model the nodes just connect to the central node. With a wireless connection, since the electromagnetic wave is omni-directional, versus unidirectional communication in the wired network, the mesh network becomes more feasible. In mesh networks with wireless communication making connections with any other entity needs just a routing protocol to support node to node direct communication. This is one of advantages of using wireless communication, but on other hand there is one challenging disadvantage which should be considered. This disadvantage refers to real-time property of some process automation applications. Fig. 1.4 is a schematic of turning wired SAN to Wireless SAN (WSAN).

Applying wireless communication in industrial automation applications introduces some problems such as real-time property of the processes which is challenging. As previously mentioned mesh topology enhances this problem as well. In digital control the sample is taken from environmental parameters at each sample period. The sampling frequency generally follows the Nyquist theorem and usually is chosen 10 times of process time constant [05Milan]. Real-time property implies that the measured value should be processed and the carrier message transferred to the actuator during the sample period before the next sample time arrives. For the fast process this sample period could be very small so that the wireless message delivery regarding to its delay in intermediate nodes and message broadcasting time delay cannot happen on time. For example a sample must be taken at every 2 millisecond, but the message transmission between two nodes takes more than two milliseconds, for instance five milliseconds because the message should be retransmitted by 10 nodes in between in order to reach to destination. This condition contradicts with real-time property. On other hand, time delay in processing the system input can lead to instability of the system [05Milan]. The usual network protocol like IEEE 802.11 and IEEE 802.15.4 are not the real-time communication protocol [04Wang], for these reason some effort is made to develop real-time wireless communication protocol for process automation like R-Fieldbus [11Lutz] or [13will]. Considering the real-time property, Autonomous WSAN (AWSAN) can be recommended for the slow processes or the processes which are supposed to be controlled by relay at present.

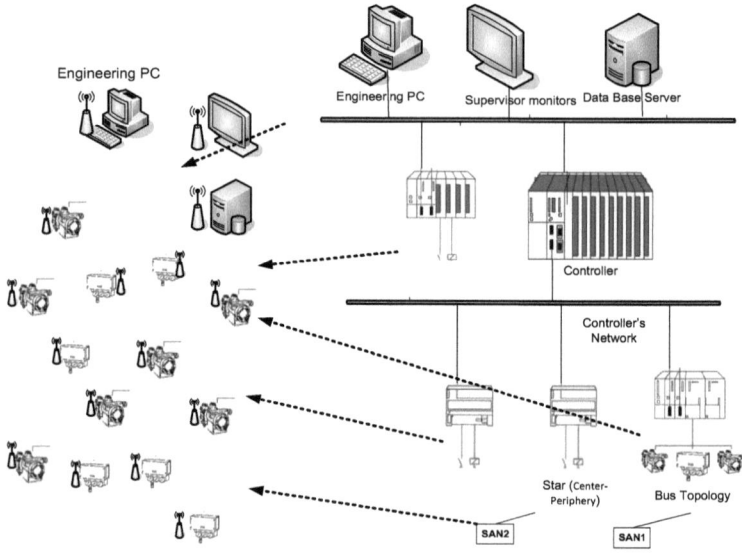

Figure 1.4: wireless SAN as alternative for wired SAN

If an automation system is a combination of different SANs distributed in a wide geographical range then each of them can be considered as an entity of an automation system. Because of geographical distribution, the entities cannot communicate with each other directly through the nodes transceiver chip because of hardware and technical limitation. Therefore each of them as an entity should be equipped with more powerful communication devices. This device which in this work is called a "courier" of the entity does not have any role in making decisions for the nodes inside entity or exchanging the data inside the entity. It is just responsible for communicating with other

1. Introduction to Autonomous system

outside entities. This courier's duty is to make decisions which are related to the whole entity, the entities role inside the system and the entities relation with other entities. From this point of view (from outside of entity) they can be called an Agent as well. An example of such a system is an intelligent container [16cont]. These types of containers are equipped with tiny wireless sensor nodes which gather environmental information and control the inner environment of the container. When they are located on the truck and moving from city to city, the small sensors are not able to exchange data with other containers or generally entities, therefore a powerful communicating node should be implemented for the trucks. As mentioned it does not and should not have interference with internal system task, so that if the courier fails the automation system of intelligent container should be able to do its job. Such an example could be seen in the BMS or HVAC applications [08CIS] as well as when a control system of a room should be aware of a condition somewhere else.

An autonomous sensor actuator network is different from what Dressler in [03Dres] calls a self-organization system. He offers two methods to distinguish the self-organized system in contrast with others. The first method is checking for existence of any kind of supervision by a supreme entity. This test result is negative in Autonomous SAN (ASAN) because in any sense there is no entity with higher authority in decision making or interconnection of nodes. Dressers' second method is checking the blue print for the system, in this sense it could be said that the ASAN follows a kind of blue print which is called a control task. In other words when a control task is defined in an automation system to control the behavior of the system, it plays a blue print role for the system in the background. Simply the automation systems are established to automate the process. This phrase is considered as a goal of an automation system. On the contrary in a self-organized system, the system behavior is an emergent property of the system [03Dres]. In fact the mutual effects of entities and their interaction and relation shape the system behavior without any external control. This concept is mentioned in another way in [01King] about decentralist federal system by anarchists.

To clarify the autonomous structure and its difference with central one the feedback control model for a process is considered. This model is shown in Fig. 1.5a, the processor block is added as representative of a decision-making unit. Separation of processor unit helps to distinguish the difference between autonomous and central WSAN- An automation system or network is combined of varieties of feedback models in Fig. 1.5a. In other words this model is the constituent element model of automation system. Fig. 1.5b depicts the part of the model that resides in center and which part in the sensor. In the central system, sensor performs measurement at each sample period. Then it sends its data in the form of a message to the center, the center compares the value with the set-point and does the processing (makes decision). Afterwards it sends its own decision as input to the actuator. In Fig. 1.5b the sensor, center and actuator tasks are separated from each other.

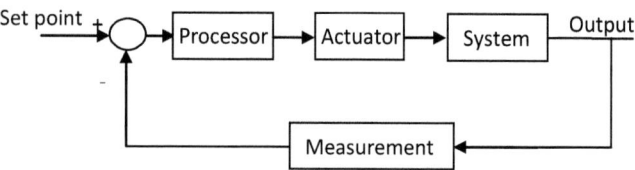

Figure 1.5a: Autonomous feedback model with elements placement

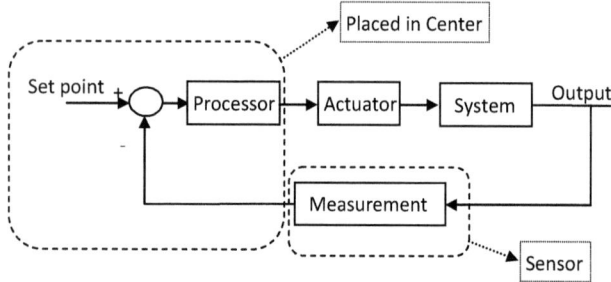

Figure 1.5b: Central feedback model with elements placement

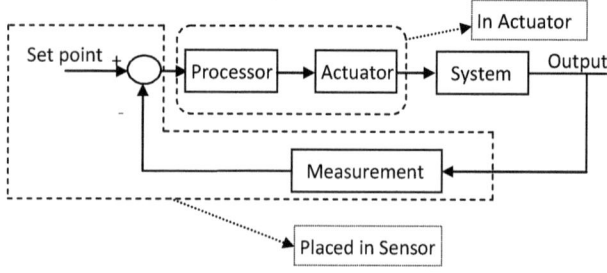

Figure 1.5c: Autonomous feedback model with elements placement

In an autonomous structure regarding what is explained about the autonomy and what is expected from distributed system, the element placement is different. In an autonomous system, a sensor performs the measurement then it does the comparison.

1. Introduction to Autonomous system

This part is illustrated by the placement of comparison part inside the sensor. Based on the result of this comparison it decides whether to send a message to actuators or not. In this manner the sensor becomes a self-decision-maker. After the message is received by the actuator, it should decide what to do. For decision making if the actuator needs information from other nodes, it fetches the data by sending a request message to other nodes. After receiving the required data the actuator makes a decision on what to do. Placing the processor element in the actuator in model Fig. 1.5c represents this decision making. In fact in autonomous system each node either actuator or sensor makes decisions for itself. The sensor decides when to send messages contrary to its task in the central system, in which the sensor sends its data periodically at each sampling period to the center. Fig. 1.5c shows which part of model is located in which node.

1.5) Logistic system

The above explained structure for WSAN can be applied to applications in a vast variety of realms of process automation e.g. logistic systems, industrial automations, HVAC and etc. In this subsection the logistic system is seen from another point of view to explain why it is needed to develop an "intelligent" entity and to see the role of Autonomous WSAN in this approach.

Suppose that there are entities A and B in an autonomous system. When they have their own resources (resources A and B) and the same goal, the competition between them appears to take control of more resources [03Dres]. This concept is depicted in Fig. 1.6. This competition can be regulated under name of cooperation protocol by sharing the resources. When one of them does not need its resource it can lend it to another entity. Such cooperation protocol makes one of the entities dependent on another one, in this way a dependent entity will be directly or indirectly subordinate to another one and its performance could be influenced by another one. In this way of system structuring there is always the potential of raising conflict and struggle between entities.

Now suppose that the entities A and B have access to one common source which makes a decision for itself to have the best service. In this case the competition between these two entities appears over the goal [03Dress]. They compete with each other to reach a better and higher goal. This is illustrated schematically in Fig. 1.7. Comparing this structure with the former one shows that the competition which could make trouble and struggle and waste the energy of the entities can be applied in a constructive way to activate the positive creativity.

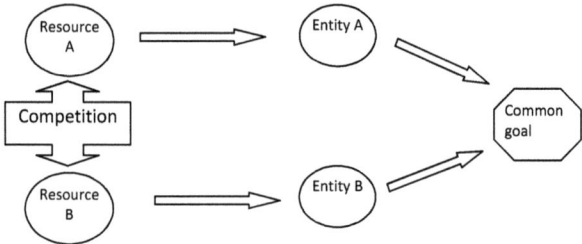

Figure 1.6: Common goal with competition over the resources.

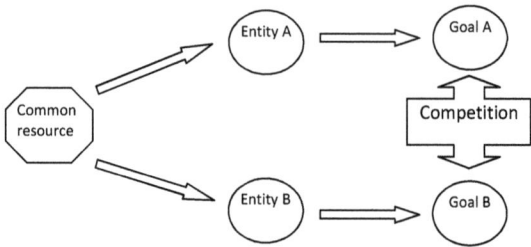

Figure 1.7: Common source with competition over goal.

To implement such a structure to the system, the resources A and B in Fig. 1.6 should be separated from the entities and appear as an independent entity inside the system. Any entity which needs the resource may send a request and the resource entity will reply to them. The new resource entity for making decisions needs to identify itself and say who it is and needs to know its own conditions. This ability to gain information, store them and use them in new situations is referred to as "intelligent" adjective.

As an empirical example suppose that entity A and B are transport companies in a logistic company. As is usual they possess some truck and containers to transport as resources. To make the containers an autonomous entity they should be made intelligent. They can be tagged with an RFID tag for identification to tell the others who they are and they should be aware of their internal situation and control their internal conditions. To achieve this aim, the autonomous wireless sensor actuator network can be applied inside the container in order to develop intelligent entity.

1. Introduction to Autonomous system

In this Chapter the Autonomous Wireless Sensor Actuator Network is conceptually introduced. In the next chapter it is explained what is needed for developing AWSAN and a sample of an applied node is introduced. In the last chapter the central and autonomous WSAN are simulated and their functionality and performance are evaluated and compared. At the last section an option is offered for network structure selection.

Chapter 2
Development of AWSAN

Abstract:
In order to develop an autonomous wireless sensor actuator network, some requirements should be fulfilled. In an autonomous network the nodes contact each other directly and for this a routing algorithm is developed which supports this feature. This algorithm is called SCAR and the first section of this chapter is devoted to the making of this routing algorithm. Firstly, according to the autonomous concepts, the demanded features are extracted, then other algorithms are reviewed to assess the features they provide. This review also helps to understand and find the constituent elements of SCAR. This routing algorithm, which uses the graph theory concepts and algorithms, is developed in follow of demanded properties in the first section. During the development of AWSAN, it was necessary to answer this question of how often the sensor should take samples from the environment. Therefore a method is proposed to find the optimal sample number. With this sample number the minimal message number and energy consumption is achieved. A specific type of wireless node is used for physical implementation and establishing AWSAN. This wireless node is introduced in the last section. Its hardware, software and functionality are introduced. This information is applied in the simulation and evaluation in the next chapter.

2.1) Routing algorithm

2.1.1) Demanded features

As it is mentioned in former section, mesh networking in AWSAN demands a routing algorithm for the network. The network structure and the application demands define the characteristic of this routing algorithm. In former chapter, it is explained that the nodes inside the autonomous network are interdependent in terms of exchanging data. They communicate with each other without involving a central node. This is the primary and distinguishable characteristic of the routing algorithm for AWSAN. The routing algorithm with which a node can communicate with another arbitrary node is called target-oriented. This classification is used to differentiate other usual types of routing protocols, which are called sink-oriented, from the routing algorithm for AWSAN. With the sink-oriented routing algorithm, one node is able to send message just to the center of the network and in more developed ones the center is also able to send message to the nodes. By this algorithm the nodes are not able to communicate with each other directly.

One of the methods for having target-oriented routing algorithm is using geographical coordinates and routing the message based on these coordinates. To provide such coordinates, a special kind of positioning system should be implemented for the nodes which could be based on embedding the Global Positioning System (GPS) or Local Positioning System (LPS). To avoid complexity, instead of adding any extra element for positioning, using the resources of the nodes can be considered.

One of the important factors which should be considered is the node energy consumption. This factor is important particularly when the nodes are supplied by limited resources i.e. small batteries. Radio transmission and reception usually consumes much greater energy than measuring (sensing) and processing data. Therefore, decreasing the transmission/reception energy consumption can have considerable impact on the life time of the network and consequently maintenance cost. The transmission and reception energy consumption is directly related to the traveling path of the message which is determined by the routing algorithm. Obviously if the routing algorithm determines longer path, the transmission energy will increase too.

Regarding to the energy consumption issue, the network topology changes should be considered as supporting features of routing algorithm as well. In general, the network topology is stationary or mobile. The routing algorithm, supporting the mobile topology, consumes energy to realize the network topology when a node wants to send a message. However, when the network topology is stationary it is not needed to consume energy for the network topology realization in time period. In industrial process automation, the applications are deterministic and usually stationary. There is a possibility that they have a moving node inside the system. Such a network can be considered as stationary even though some nodes are mobile. In other words the wireless nodes can always move inside the radio range of their corresponding neighbors without considering the network topology as a mobile network. As example, HVAC applications and intelligent containers in logistic systems [16cont] can be named. Although the container is moving on the track, but the inner topology of the container is relatively stationary. Regarding the aforementioned points, the routing algorithm for AWSAN is supposed to support stationary network.

The AWSAN is introduced for applying distributed system over the SAL to get the advantages of distributed system such as scalability. Therefore, the routing algorithm not only should not contradict with such properties but also it should support them. Then consequently one of other characters, which are demanded from the routing algorithm, is the scalability. The question is that if the application demands to add new nodes, would the routing algorithm allow this addition to happen easily? Some routing algorithms are table driven, that is, each node has a table of the intermediate nodes addresses or IDs on the path to the destination. Therefore, if a new node is added to the network, the new table should be defined for all nodes which are connected to this node. On the other hand when the size of the network increases other problems can appear. This side problem refers to this fact that much more memory will be required to save the table inside the nodes. In this sense scalability faces with constrains. These kinds of routing algorithms are classified under the category of "proactive" algorithms.

Contrary to the proactive algorithms there is another category of algorithms called "reactive" algorithms. In reactive algorithm the path is not already known and is computed every time a message should be routed. The computation is done based on the node's stored information and information carried in the message. These kinds of routing algorithms deal with scalability much easier than proactive algorithms. Some routing protocols combine the properties/characteristics of reactive and proactive algorithms to support the mobile network more effectively. However, as it was already mentioned, the mobility costs/consumes more energy which is not necessarily useful for AWSAN. We considered two basic categories: proactive and reactive algorithms. Regarding to this categories, AWSAN should be reactive algorithms. One important factor for reactive algorithms is the processing power of the processor of the nodes. The point is that the next node computation should be time consuming for the processor of the node. If this happens then the nodes capacity for further computations will be

2. Development of AWSAN

weaken and the capability of offering routing service to received messages will be decreased.

By now it is clarified the below listed characters for the routing algorithm in AWSAN are required:

- Target-Oriented
- Without embedding element
- Energy consumption consideration
- Supporting stationary network
- Scalability
- Reactive algorithm
- Easy computation

In this research the routing "Protocol" and outing "Algorithm" are discriminated. The protocol in this paper represents the layers of 7 layer ISO model but what is meant about the algorithm is just network layer of ISO model, in which next node should be found. In this way the effort is focused on the essential element of finding the next node or path. The problems such as how the nodes deal with dead neighbors is left for developing a routing algorithm to the routing protocol.

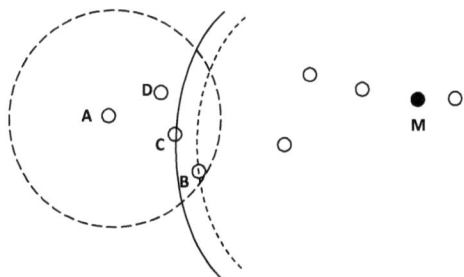

Figure 2.1: Greedy forwarding example

In some routing algorithm there is one problem which puts reliability of the algorithm under question. This problem is known in different terms such as "Void" [17gpsr][18lcr] or hole [19gear]. This problem happens when the routing algorithm logic fails to response. For example suppose that a constituent element of a routing algorithm is "Greedy" component. Based on the greedy definition, always the nearest neighborhood to the destination will be chosen. With this logic it may happens that there is no nearest neighborhood to the destination but there is still a path to the destination. To explain this situation, the example in Fig. 2.1 is studied. In this situation the greedy algorithm logic dictates that the node A chooses node B and next intermediate node. But as it could be seen in Fig. 2.2 there is no nearest node to the destination after node A whereas there is still path to destination. Routing algorithms deal with such situation with different methods. In order to develop a routing algorithm for AWSAN firstly the possibility of the void problem existence should be verified and secondly a method should be developed to deal with the Void problem. In general it could be claim that existing void problem is considered as negative point because it increases the core of programming and computation and increases complexity.

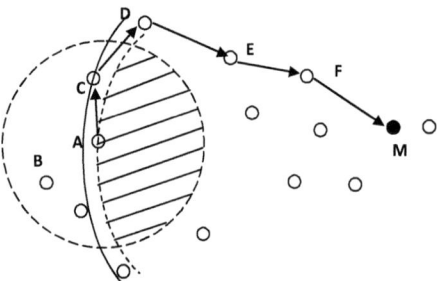

Figure 2.2: Void example and perimeter solution

2.1.2) Literature review

Before developing any routing algorithm, at first the existing routing algorithm are reviewed to find proper routing algorithm. Because of central thinking pattern domination on the SANs, the majority routing algorithms follow the same pattern. It means that most of routing algorithms are sink-oriented and they are supposed to be used for monitoring the environment not controlling them, therefore being target-oriented becomes important in evaluation of the routing algorithms and becomes motivation to develop a new routing algorithm.

Greedy Perimeter Stateless Routing (GPSR) [17gpsr] is a routing algorithm for wireless networks over IEEE 802.11 and it is not designed for IEEE 802.15.4 (LR-PAN) network, consequently it does not concern the energy consumption of the network. But since it uses the greedy component and it could be base for other algorithms. In this algorithm the nodes are equipped with positioning system. The nodes know about their neighbor geographical position but not about the entire network topology. This is the reason it is called stateless and this fact implies that the GPSR is reactive algorithm.

2. Development of AWSAN

For finding the next node the algorithm includes two methods: Greedy and Perimeter. If Greedy method faces with Void problem, the perimeter method comes to service point. In perimeter algorithm the nodes uses either Relative Neighborhood Graph (RNG) or Gabriel Graph (GG) for realizing network graph then based on the right hand rule the next node will be chosen until it goes out of void situation. Again here the nodes do not calculate the graph for whole the network, they just computes the graph for their neighbors. GPSR can support stationary and mobile network. Depends on the mobility rate of the network topology the rate of exchanging the information between the neighbors will change. GPSR is categorized under target-oriented algorithms.

Yan and etl in [19gear] propose Geographic and Energy Aware Routing (GEAR). This routing algorithm is based on the geographic coordinates; therefore the nodes should be equipped with positioning elements. This routing algorithm is designed for stationary network. At first step it forwards the message to a region then by one of two mechanisms: recursive geographic forwarding or restricted flooding, the message will be disseminated inside the region. The nodes not only aware of their neighbors coordinates they are aware of their energy level also, then when they are making decision for choosing next forwarding node they consider the energy level of the nodes too and in this way the network energy consumption will be distributed over the nodes more evenly.

Other routing algorithms such as LEACH [20leac], ZRP [21zrp] and AODV [22aodv] are studied in this work too. LEACH is designed for communicating between the clusters in the network and it is sink-oriented. This routing algorithm neither its defined application nor its property does not suit for AWSAN. The ZRP and the AODV are designed for mobile network and they are combined of proactive and reactive algorithm. They pay cost to support the mobility of the network which is not necessarily for AWSAN.

GRAdient [23grad] routing is one of other routing algorithms which is assessed for AWSAN. The first property attracts the attention is that it is a sink-oriented algorithm which is not proper for AWSAN, but the basic idea behind this routing algorithm is interesting for further development. This algorithm is self-configurable; it means that after deploying the nodes they start to figure up the routing element requirement. Therefore at first step the network enters the set-up phase then it goes to operational phase. In set-up phase the sink starts to broadcast a message, the neighbors forward the received messages and then the other nodes do the same. In this way the cost field shapes and all nodes are labeled with cost value. In operational phase when they receive a message they compare the value carried inside the message with the nodes label and regarding to the gradient value they realize that the message is forwarding or back warding. The nodes label is called "Cost" which for example could be the minimum required energy to broadcast a message (broadcast cost), the minimum required energy to transmit the message (transmission cost), the shortest time for reaching to the sink, the minimum hops number or combination of the different parameters. GRAdient shapes the cost field based on the energy consumption. In order to do this, it sends the energy level of the message with the message and on another side the receiver measures the received message energy level and then it calculates the dissipated energy. In order to avoid facing with the void problem it uses the credit label.

In general the basic idea of cost field is widely spread in different routing algorithms, for example the minimum hop routing algorithm [29mhr] shapes the cost field based on the hops number. For AWSAN application the problem concerning to GRAdient is the sink-oriented characteristic of the algorithm. Since the label of the nodes is one dimensional, obviously the message is forwarded just in one direction. As it is mentioned, target-oriented property of a routing algorithm means that a message should be able to goes around in three dimensions. This implies that label of the nodes should be at least 3 dimensional in order to support three dimensional message forwarding. Such idea is followed in Logical Coordinate Routing (LCR) [18lcr] and in [32anto][33fang]. In these algorithms the sink is called as Land mark node or anchor, because they are supposed to be reference node for labeling and establishing cost field. They take at least three anchors (resembling the geographical X, Y and Z axes) in the network and they repeat the same procedure as GRAdient for each sink to set up cost field. As result each node will be labeled referring to each anchor and the cost field would have more than one dimension. For routing a message from one node to another one they rely on the coordinates of the node and follow the gradients of the nodes or they use greedy algorithm to find logically closest node to destination. But there is one important point here which is not usually noticed. The point is that even the nodes have unique coordinates but the gradient is not necessarily unique, thus different nodes are possible to receive the message or in other words the message can be routed in different paths. On other hand for the algorithms with greedy component it is possible that two nodes locate logically in the same distance of a sending node.

The Sequential Coordinates Routing Algorithm (SCAR) [34sca1][35sca2] is proposed for AWSAN which follows the similar idea. The difference is that the coordinate bases are not the anchors. The coordinates assigned based on the arrangement of the nodes over the Minimum Length Tree (MLT) of the network graph. The coordinate space dimension depends on the maximum number of edges connected to at least one of the nodes which is referred as node degree and on the nodes arrangement. Using MLT for routing is already developed in other way in MSP [31msp] protocol in Cisco routers. Briefly by

SCAR, the network graph is extracted and then the unique coordinates are assigned to the nodes starting from arbitrary node. The message header carries just the destination and next receiver node coordinates. In each node, next node coordinates is computed and the message will be forwarded to it. Moving along the tree prevent of having any loop in forwarding message and facing with the void problem. MLT in this routing algorithm generally leads to minimum energy consumption in comparison to other trees. This algorithm is self-configurable therefore like GRAdient it has two phases: Set-up phase and operational phase. Both phases and different characters of SCAR are explained in follow.

2.1.3) SCAR development

A) Setting up phase

For set-up phase one node is chosen to do the computation. This node is called base node which could be a usual node, a hand held device or the courier of a network which is explained in former chapter. After deployment of the nodes the set-up phase starts. The basic idea is that all node start to broadcast a "Hello" message with maximum transmission power. This message contains the senders assigned ID. Any node in neighborhood which receives the "Hello" message, records the signal strength with the senders ID, then it sends a message back to the sender containing this information. In this way each node can make table of its neighbors with the Channel Attenuation Value (CAV). The CAV is calculated from difference between the transmitted signal power and received signal strength. After finishing the nodes negotiation part, the nodes start to send their CAV table to the base node. This monitoring task can be done by simple flooding. Before explaining what the base node does with the collected data from the network, a practical issue for nodes negotiation should be explained.

Practical Implementation strategy for negotiation

There are two situations in which the message can be lost during the set-up phase. The message lost during this phase is important because can lead to deform the network graph and lose the connectivity of the nodes. These two situations can happen during nodes negotiation by "Hello" message and during transferring CAV table to the base node. To cope with this issue, two methods are proposed for implementation. Both of them basically functions with the minimum hop routing [29mhr] algorithm which is sink oriented.

The first method is simple flooding. The negotiation starts from base node. After the neighbors received the message, they wait for a random time and then reply to the base node and broad cast for their neighbors too. The neighbors could be in next level, level 2 in Fig. 2.3, or be in the same level (level 1). The same procedure can be done for other levels. In each step some nodes play the role of the base node and some nodes plays as receiver and in next step the same nodes play the role of the base node. This procedure follows to the last level. In order to compensate any message lost, the procedure can be repeated from beginning for a few times and the average of CAV table be taken for further evaluation.

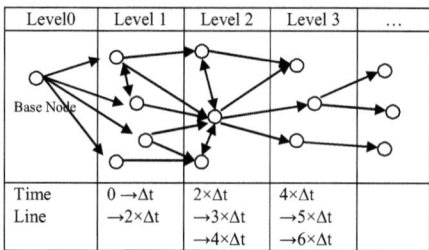

Figure 2.3: Slotted flooding for set-up phase

The second method called "slotted flooding" which is inspired from Slotted Aloha [24tanb]. The basic idea of this method is separation of the receiving message phase from broadcasting phase for the next level. In this method two time intervals with the length of Δt are devoted to nodes in each layer. The time line starts when the base node starts to broadcast or more precisely when the nodes in level 1 receive the message from base node (Fig. 2.3). The nodes in level 1 broadcast randomly within [0 Δt]. The message contains the broadcast timestamp denoted by t_1. On one hand the base node receives this message, and then it can make its own CAV table. On other hand suppose that a node in level 2 receive the message with timestamps t_1, then the nodes waits for Δt- t_1. In a random time within the next time slot [Δt 2×Δt] the node replies to the level 1 and they will make their own CAV table. After beginning the next time slot [2×Δt 3×Δt] the nodes in level 2 will broadcast for nodes in level 3 randomly. The nodes in level 3 in time slot [3×Δt 4×Δt] play the same role of the nodes in level 1 in time slot [Δt 2×Δt]. This procedure continues to the last level. In slotted flooding method, replying to higher level and broad casting for lower level are separated in distinct time slots which leads to reduction of the data lost probability.

The time slot Δt is chosen based on the packet broadcasting time and estimated maximum received message number by a node in a time slot. The first factor depends on the radio chip bit rate and the second one depends on the density of the network. Suppose that the network support IEEE 802.15.4 protocol [30154] and the radio chip bit rate is 250 kbps. With the packet length equal to 40 bytes in TinyOs [37tiny], the time for reception or sending a message will be equal to 40×8×1/250000=1.28 ms. For example if a node has almost "m" neighbors in average the time slot could be taken Δt=n×1.28 ms so that "n>>m". It needs "m" independent 1.28 millisecond time slot during each Δt. The probability of the transmission or reception overlapping can be calculated based on these factors. By increasing Δt, the probability of collusion can reduce. On another point of view changing Δt makes it possible to adapt this method to the network size. In practical implementation this method presented that it is more reliable especially when the size of network increases and needs less set up time. The data fusion mechanism [25fusi] can be applied to decrease the energy consumption and message traffic.

Extract the MLT of the network

After this stage all nodes send their CAV tables to the base node. The base node realizes the network topology by analyzing these tables and making adjacency list or matrix [26gary]. This graph is made of vertices, edges and edges weight correspondently representing the nodes, nodes connectivity and channel attenuation. In set-up phase if any node is isolated and does not have any connection to any other nodes it will not send and receive "Hello" message and its table will not exist in base. Likewise the same happens for a group of nodes, therefore the constructed network graph in base is always connected graph. A connected graph is graph in which from each node there is at least one path to other nodes [28jung]. It is assumed that there are no isolated nodes in the network because they will be useless and not counted as part of the network.

After shaping realizing the network graph, the base extracts the MLT of the network graph by "Kruskal" algorithm [28jung]. The messages are supposed to be routed over this "Tree" thereby the routing will be without cycle and void problem because the "Tree" as defined in [28jung] is a connected graph without any cycle, it means that from one node to another one there is just one path. In Fig. 2.4 a hypothetical network graph is illustrated by the dashed edges and its MLT is shown by the thick edges.

The MLT is not the only tree of the network graph. The other trees could be found as well but focusing on the MLT returns to the relation between the network energy consumption with routing algorithm functionality in AWSAN. It is claimed that by MLT the network energy consumption would be less and message delivery would be more reliable. The basic explanation for the Kruskal algorithm is that the edges are sorted from the lightest weight to the heaviest, then it starts to pick up from the lightest edge one by one to form new edge set. If choosing and adding an edge from graph edge set to the new edge set causes cycle creation in new edge set, it will be neglected. In this algorithm the priority is with the edges of lightest weight. In base node the weights of the network graph edges are channel attenuation value. It infers that when we are choosing the MLT, the priority is given to the channels with less signal strength attenuation. This way of selection has two advantages. The first one is that the more reliable connections are chosen. By considering the fact that the transceiver chip supports the programmable output (explained in section 2.3), the second advantage is that the less signal strength fading provides the option to reduce the broadcasting power and consume less energy for transmission.

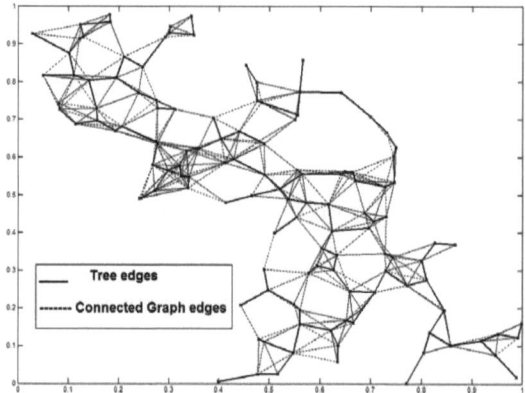

Figure 2.4: A sample connected graph with its MLT

Extract the Trunk and assigning the sequential coordinate

After MLT is extracted, the trunk of the MLT should be realized. The trunk is the path from base node to the node with degree one and maximum hops number. The node degree is defined as number of edges connected to a node [26gary]. The trunk is realized by the Deep First Search (DFS) algorithm [27foul]. The nodes on the trunk are called root nodes. The root nodes are connected to the remained nodes with third or more degree. The trunk extraction algorithm is applied to extract the limb and the same is repeated over limb to extract the branches and twigs.

The sequential coordinates are in the format of $(\alpha_1,...,\alpha_n)$. For the root nodes α_1 is the sequence number from base node and the rest are zero. For the nodes on the limbs α_1 is the same as root node coordinates value and α_2 is equal to sequence number from the root node on the limbs and the rest coordinates are zero. If a root node degree is m and

2. Development of AWSAN

greater than 3, then m-2 coordinates element i.e. from α_2 to α_{m-2} are dedicated to the limbs and the element corresponding to each limb has its value while the rest of the dedicated elements are zero. The same procedure for assigning the coordinate on brunches and twigs is repeated. In this way all the nodes will have a unique coordinates and based on these coordinates the message will be routed. Fig. 2.5 depicts an MLT, trunk, limbs, branches and twigs for above example network graph in Fig. 2.4. It also shows some of assigned coordinates to the nodes. The maximum coordinates dimension depends on the highest node degree and their arrangement.

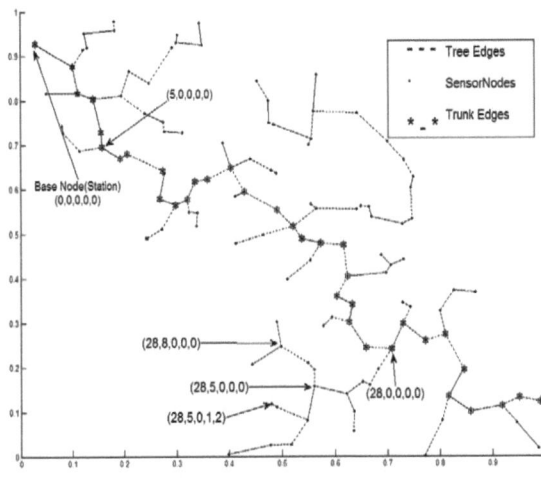

Figure 2.5: A MLT with sequential coordinates

By this procedure and algorithms the base node generates the coordinates for the nodes. Afterwards it starts to distribute the coordinates between the nodes. Now it has network graph and can send designated message to each node to deliver coordinates or it can just flood a messages with a table of the nodes ID and their coordinates. But because of the reliability issue in message delivery, message flooding by base node is desirable for implementation. The nodes after receiving the message they look up for their own coordinates.

B) Operation phase

After the nodes received their coordinates the network can enter into the Operation phase. In this phase whenever a message is received three states happens. If the message was destined to that node, the message traveling will finished there. If the destination coordinates is not the same as receiver nod but the next node coordinate inside the message header is the same as the receiver nodes coordinates then the node starts to calculate the next forwarding node coordinates with a simple routing procedure shown in Fig. 2.6. After computation the next node coordinates it insert the computed coordinates into message header and broadcast the message. When a node receives this message it follows the same procedure. If a node receives a message which neither its destination was that node nor its coordinates were in the message header as next forwarding node, then the message will be dropped.

```
1) (c₁,...,cₙ) = (a₁,...,aₙ);
2) for      i=1 : n
3)     if  aᵢ ~= bᵢ
4)         for    j=n : i+1
5)             If          aᵢ ~= 0;
6)                         cᵢ = aᵢ - 1;
7)                         break;
8)             end
9)         end
10)        cᵢ = aᵢ + sign(bᵢ-aᵢ);
11)    end
12) end
```

Figure 2.6: SCAR procedure

To explain how the algorithm in Fig. 2.6 works, suppose that a node with coordinates $\underline{a}=(a_1,...,a_n)$ receives a message. The message header contains the destination coordinates $\underline{b}=(b_1,...,b_n)$ and next forwarding node coordinates $\underline{c}=(c_1,...,c_n)$. Considering most complex case we suppose that the destination is on another limb. In this case line 6 forwards the message over the sender node limb and branches to its corresponding root node on the trunk. Then line 10 transfers the message over the trunk to the destination limb. Afterwards the message will be forwarded over the destination limb to the destination node. The direction of forwarding or backwarding is distinguished by gradient of destination coordinates and current node coordinates.

C) SCAR evaluation

At the beginning of the routing algorithm section, the demanded features from routing algorithm for AWSAN are discussed and counted. Now the SCAR should be evaluated based on those features. The foremost feature is target-oriented. SCAR is target-oriented because any node is able to send message to any arbitrary destination. This arbitrary destination can be center in central network as well. This means that SCAR is not only target oriented it is sink oriented as well. Thereby in next sections to compare Autonomous with central network SCAR is used for both networks. This routing protocol does not need any positioning system like GPS. The broadcasting power is known and the nodes should just measure the received signal strength. The radio chips, which support IEEE 802.15.4, have a registry to save the Received Signal Strength Indicator (RSSI) [30154]. For SCAR this registry value could be used for the CAV, therefore it does not need further embedded element.

As the core of routing algorithm in Fig. 2.6 shows, SCAR is reactive algorithm because the next node coordinates is found by calculation. The core program has very easy computation therefore it could even be integrated to very limited microcontroller and it does not occupy the microcontroller for long time. In terms of scalability, when a node should be added to the network the node starts to send "Hello" message and after receiving the answer, it chooses the neighbor with the smallest CAV then asks its coordinates. It sets up its coordinate based on its neighbor one. It can be done by appending one coordinate. During the SCAR development explanation, it is explained how selecting MLT as layout is related to energy consumption consideration and it provides more reliability for message delivery. On another hand it does not face with void problem which leads to increase the complexity of the network. SCAR is designed for stationary network, although the nodes inside of the radio range of their corresponding neighbor(s) can move but the topology of the network is stationary.

2.2) Sample number

2.2.1) Sampling theory

Implementation of wireless communication in industrial automation faces with challenges. One of the challenges appears is the real-time property of the system. In digital control system, the control variables are sampled and then the values are processed. The sampling period depends on the system time constant which defines the natural frequency of the system. Nyquist-Schannon sampling theorem states that the sampling frequency should be greater than double of the system natural frequency, otherwise in frequency domain the sampled signal frequency spectrum will be shaped by overlapping the continuous signal frequency spectrum. In this way sampled signal does not represent the behavior of the continuous signal and it will not be reconstructable [41open]. Ideally the sample frequency is 10 times larger [05Milan].

When a sample is taken the sampled variable should be transmitted and processed before next sample arrives. If the data cannot be processed and control task cannot be achieved during the sample intervals, it means that the control system cannot follow the system variations and it does not hold the real time property. Such phenomena can lead to instability [19]. From another point of view if the Z-transform of the system model is considered, the sampling period has impact on the pole and zero of the system model and consequently the system behavior changes. By higher sample frequency the sampled signal resembles the continuous signal form more and the frequency spectrum will have better resolution. In wired network the sample frequency can be increased as high as enough. In industry automation applications, Programmable Logic Controller (PLC) is used as controller presented in Fig. 1.3. The PLC takes samples from signals at its input ports and then executes the control program, performs the computations. After all it finally updates the output ports. This procedure is presented in Fig 2.7. The procedure is repeated cyclically. The cycle time is monitored by a watchdog timer. If the cycle time increases for any reason e.g. male function of the program, watchdog timer declares an error signal and may restart the cycle again. On other words the input signals are scanned with the cycle time period, it means that the cycle time plays the sample period role for the control variables. In order to cover every process even fast one and keep being real-time the cycle time is taken very small.

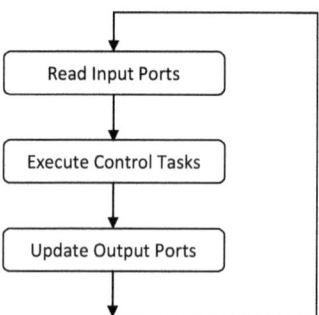

Figure 2.7: PLC functionality procedure

High sample frequency has some consequences in wireless network in comparison to the wired network. These consequences should be analyzed and dealt with, because they are directly related to the efficiency of the network. Higher sample frequency causes more message transmission in wireless network. Higher message transmission number means higher network traffic and it leads to higher delay time in message transmission and more network energy consumption. This energy consumption factor becomes more important when the nodes are supplied by batteries, because it will mean that the battery will depleted faster and maintenance cost will increase. On the other hand as it is mentioned in above paragraph, delay time in transmission can lead to instability of the system and deteriorate the control quality of the system. Therefore the sample number cannot be increased the same as it is done in wired network. In this sense the sample number is compromised with the transmission number or nodes energy consumption which is explained in follow. The proposed method which is explained in this section is already published in the international conference and journal, [49opt] and [52opt].

2.2.2) Actuator frequency

For computation of optimal sample number the system feedback model in Fig. 1.5a is taken into consideration. The basic assumption for this section is that the system is linear time invariant, therefore it has Laplace transform. The transfer function of the system is first order and the system is assumed to be slow enough so that it can be controlled with AWSAN. The actuator of the system is on-off relay. For example computations some parameters are assumed to be known. These parameters are just examples for clearer explanation. The method is general but the computed values are not necessarily valid for each application.

$$H(s) = 1/(T_n \times s + 1)$$ Equation 2.1

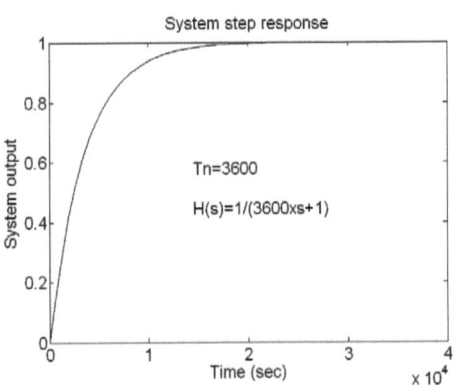

Figure 2.8: System step response

It is assumed that the transfer function is first order and its Laplace transform is presented by H(s) in Eq. 2.1. Fig. 2.8 depicts the step response of this system with T_n=3600 s. The set point value is taken as Y_0 and the limit values are Y_{hc} and Y_{lc}. The limit values distances from set point value (Y_0) are equal. Fig. 2.9 shows the system step response with relay controller in ten hours. The instance limits value are chosen as Y_{hc}=0.7 and Y_{lc}=0.5. The actuator On-off frequency in a continuous domain is calculated by Eq. 2.2.

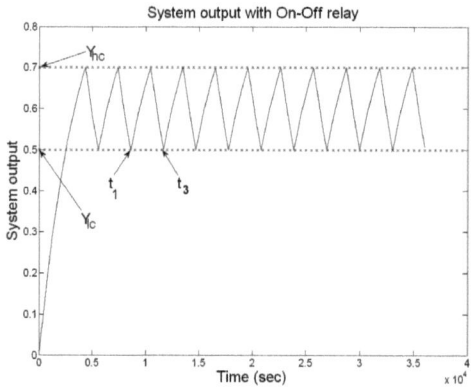

Figure 2.9: System output with relay controller

$$f_c = \frac{1}{t_3 - t_1} = \frac{1}{T_n \times \ln(\frac{Y_{hc} \times (1-Y_{lc})}{Y_{lc} \times (1-Y_{hc})})} = \frac{f_n}{c}$$

Equation 2.2

$$y(n) = (1 - \exp(-T_s/T_n)) + \exp(-T_s/T_n) \times y(n-1)$$
$$T_c = N \times T_s$$
$$f_s = N \times f_c = (N \times f_n)/c$$
$$y(n) = (1 - \exp(-c/N)) + \exp(-c/N) \times y(n-1)$$

Equation 2.3

The system is still in continuous domain. The actuator oscillation frequency is calculated and presented in Eq. 2.2. It can be observed that the actuator frequency is a function of system natural frequency divided by a coefficient. This coefficient is function of the limits values. For digital calculation and digital control, the system model should be transformed to discrete domain. In discrete domain the system output is sampled at

2. Development of AWSAN

each T_s time interval. Sample number (N) is defined as division of actuator oscillation period (T_c) over the sample interval (T_s) (Eq. 2.3). The final aim is finding either the sample number (N) or sample period (T_s). After mapping transfer function to the Z-plane, the recursive equation is calculated. Eq. 2.3 shows the recursive equation for a normalized output of above system. The recursive equation is rewritten in Eq. 2.3 to show its dependency on the sample number (N).

2.2.3) Discrete control limits and lower limit boundary for sample number selection

By moving to the discrete control domain, the control system faces with a problem. Suppose that the last sample is taken just before the limits (Fig. 2.10), then until the next sample the relay does not realize that it should switch to another state. Subsequently the system output can go beyond the limits. As example suppose that we are going to control the temperature of a room about 20° C. The limits are supposed to be 18°C and 22°C. With this phenomenon it can happen that the temperature oscillates in broader range which is not acceptable. This range is strongly depends on the sampling interval. Larger interval leads to broader range oscillation. This incident is considered as an error.

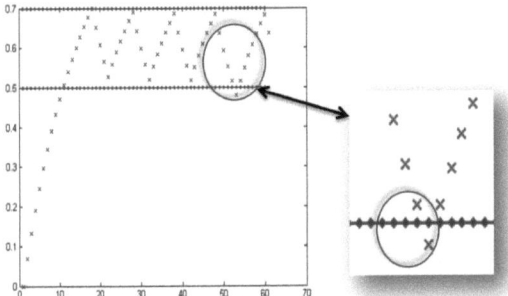

Figure 2.10: Error caused by sampling

With the intention of stepping away from such errors, the new limit values are defined in discrete time system. The maximum error happens when the former sample is taken just with very small distance from the limits. In this case the next sample will happen with maximum distance from the limits. To set up the new limits, it is assumed that samples are taken on the higher and lower limits then one step is taken backward and the former sample is taken as discrete limits value. By this choice even if the worst case in discrete domain happens and the sample is taken very close to the discrete limits, the next sample is still inside the continuous limits band. These limits are called Y_{hd} and Y_{ld} in discrete domain and Eq. 2.4 shows how they can be computed from the limits and sample number. Obviously the interval between the limits in discrete domain is smaller than same interval in continuous domain; therefore actuator frequency in discrete domain (f_d) is greater than f_c. Regarding to Eq. 2.4, lower sample number causes the difference between the limits value in discrete and continuous domain, i.e. (Y_{hc} − Y_{hd}) and (Y_{lc} − Y_{ld}), increases, therefore the limits band width in the discrete domain becomes smaller and the actuator frequency increases. This consequence should be dealt with.

$$Y_{hd} = (Y_{hc} + \exp(-c/N) - 1)/\exp(-c/N)$$
$$Y_{ld} = Y_{lc}/\exp(-c/N)$$

Equation 2.4

$$N > c/(-\ln(1 + Y_{lc} - Y_{hc}))$$

Equation 2.5

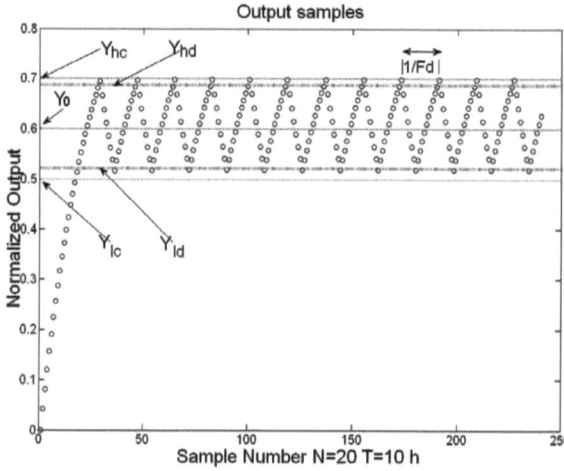

Figure 2.11: Digitized system output

Since the sample number has impact on the discrete limits, obviously the sample number should not be selected so that the higher limits become smaller than the lower limit value, which means $Y_{hd} > Y_{ld}$. This inequality leads to lower boundary for sample number stated in Eq. 2.5. This is the first criteria for choosing sampling number. As an example for Y_{hc}=0.7 and Y_{lc}=0.5, N should be strictly greater than 3 regarding to Eq 2. 3 the sample frequency should be at least 4 times of actuator frequency $f_s \geq 4*f_c$. In Fig. 2.11, the digitized output for the above system with N=20 is depicted. In this figure if the actuator status changes are counted and actuator frequency is computed, it can be observed that its frequency is increased 20 percent versus its value in the continuous domain.

2.2.4) Actuator frequency drift

In above section it has been explained how determining new limits leads to the actuator frequency increase; we call this increase as actuator frequency drift. The normalized actuator frequency drift is calculated in Eq. 2.6. As it is already explained and Eq. 2.6 shows the actuator frequency drift is proportional to the sample number. By lower sample number, the actuator frequency drift increases and by higher it decreases and gets closer to its value in continuous domain ($f_d \rightarrow f_c$). At first look this drift is important because it has impact on the actuator life time but apart from that it increases the message number as well.

It has already mentioned that in AWSAN the sensor sends message to actuator when the system output reaches its limits. By looking at Fig. 2.11 it can be seen that the message number is double of the actuator status changes number (i.e. one message for on-off and one message for off-on transient states). It infers that the message number is directly proportional to the actuator frequency. Reduction of the sample number causes the actuator frequency rising which leads to the message number rise up and it is not desired. Therefore it is better to increase the sample number but it can cause side problems as well. By raising the sample number, the occupancy and energy consumption of the node processor will increase. The increase of processor occupancy reduces its service for routing task in mesh network and it leads to more messages lost. By consideration the consequences of both sides of increase and decrease of sample number, the optimal sample number should be extracted. The optimal sample number for AWSAN is defined so that neither to be so small that causes the extreme high actuator frequency drift and message number nor so large that the processor becomes too occupied and the process energy consumption increases highly.

$$P(N) = \Delta f / f_c = (f_d - f_c)/f_c =$$
$$\frac{c}{\ln(\frac{(Y_{hc} + \exp(-c/N) - 1)(\exp(-c/N) - Y_{lc})}{Y_{lc} \times (1 - Y_{hc})})} - 1 \quad \text{Equation 2.6}$$

In a CWSAN, sensor sends message to the center at each sample time and the center sends message to the actuator when the measured parameter reaches the limits. In this structure each sample point represents one message transmission from the sensor to the actuator. Therefore higher sample number leads to higher message number which causes more transmission energy consumption and high network traffic. On other hand reduction of the sample number leads to the rising of the actuator frequency whereas it is not good for actuator life time and the center should send more messages to the actuator.

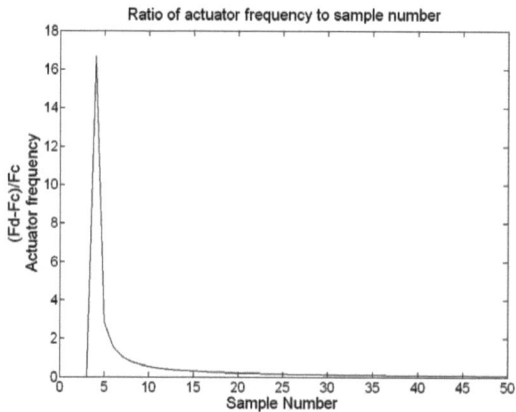

Figure 2.12: An example of actuator frequency ratio

In order to have clearer understanding of the actuator frequency drift behavior, the normalized drift is plotted versus sample number in Fig. 2.12 for the parameters $Y_{hc}=0.7$, $Y_{lc}=0.5$ and $T_n=3600s$. Based on the Eq 2.5 the lower limit for sample number is computed and it should be greater than three (N>3). From Fig. 2.12 it can be seen that for N=4 the actuator frequency jumps up 16.67 times (1667 percent). On one hand this jump indicates about 16 times more messages should be sent to the actuator and on other hand this oscillation for actuator is not reasonable either. The solution for resolving this situation is increasing sample number. Suppose N=20, then the actuator frequency drift will be about 20 percent which is more acceptable considering the former drift with N=4. If we continue to increase the sample number from N=30 to N=50, the actuator frequency decreases just about 6.5% but this increase (from N=30 to N=50) causes 66.66% rise of process energy consumption, 66.66% rise of the node processor occupancy in AWSAN and moreover it causes the same percent rise of message number in CWSAN. This increase sounds not useful and reasonable, in this

2. Development of AWSAN

sense the sample number increase can be cut off in compromising with actuator oscillation.

2.2.5) Actuator frequency drift versus limits interval

Similar to compromising the actuator frequency drift with sample number, the same can be done with the limits interval. Assume that the system designer has choice for upper and lower limits value, so that the set point value Y_0 is given and the limit distances (ΔY) from Y_0 are equal but not fixed. Revising Eq. 2.6 with ΔY and Y_0 results in Eq. 2.7. In this equation Δf is function of (N, ΔY) and one parameter can be derived by two other parameters. As example we suppose 20 samples number (N=20) and maximum 20 percent actuator frequency drift (($\Delta f / f_c$) ≤ 0.2) are acceptable for the sensor and actuator, therefore ΔY based on the computation should be 0.18.

$$\Delta f / f_c = (f_d - f_c)/f_c =$$

$$\frac{c}{\ln(\frac{(Y_0 + \Delta y + \exp(-c/N) - 1)(\exp(-c/N) - Y_0 + \Delta y)}{(Y_0 - \Delta y) \times (1 - Y_0 - \Delta y)})} - 1 \qquad \text{Equation 2.7}$$

$$c = \ln(((Y_0 + \Delta y) \times (1 - Y_0 + \Delta y))/((Y_0 + \Delta y) \times (1 - Y_0 - \Delta y)))$$

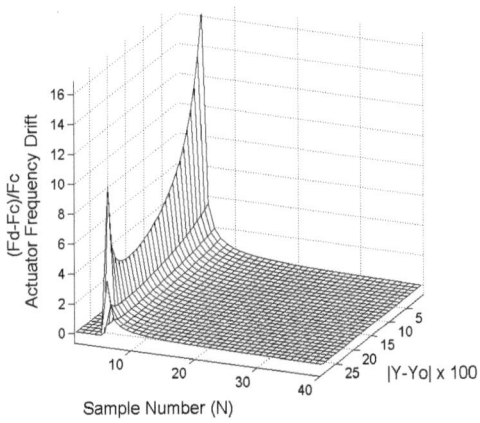

Figure 2.13: A sample of actuator frequency ratio

To see the behavior of the Eq. 2.7, it has been plotted in Fig. 2.13 with Y_0=0.7. This figure illustrates that for high sample number changing the limit distances (ΔY) to the maximum possible values, the actuator frequency reduces about 50 percent. On the other hand, for small sample numbers, larger limits band width (2xΔY) does not necessarily lead to a lower actuator frequency drift. Eq. 2.7 provides a compromising condition between three parameters: the actuator frequency, the limits and the sample number.

2.2.6) Sample number selection in CWSAN

The schematic structure of sensor–actuator communication in central network is depicted in Fig. 2.14. As it is already introduced in Fig. 1.5b, the sensor measures the environment parameter and sends the value at each sample period to the center. The center compares the value with the limit values, in case that the value sent by sensor is beyond them, the center sends a message to the actuator to inform the actuator. In Fig 2.14 "hop" represents repetitions number of message. If the message is sent directly to the center by sensor without any intermediate node involvement, then hops number is equal to one (r=1). If there is one intermediate node between sensor and center which involved in message transmission then r=2. In general, "r" or "s" is equal to the number of involved nodes in message transmission plus one.

In order to calculate the message number in Fig. 2.14, it is supposed the sample number is equal to N. The message number from the sensor to the center during time T is equal to $T/T_s \times r = ((T \times N)/T_c) \times r$ and the message number from center to the actuator equal to $(T/T_d) \times 2 \times s$. T_d in this relation is actuator period in discrete domain. If its equivalent from Eq. 2.6 is substituted, $(T \times (p(N)+1)/T_c) \times 2 \times s$ is derived. Total message number is summation of these two values. The total message number in time unit is expressed in Eq. 2. 8.

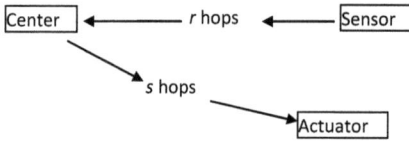

Figure 2.14: Central network structure

$$g(N,r,s) = (N \times r + (p(N)+1) \times 2 \times s) \times f_c \qquad \text{Equation 2.8}$$

2. Development of AWSAN

Now the message number can be calculated for each pair of sensor and actuator nodes. The sample number in Eq. 2.8 can be chosen so that the function output which is message number becomes minimal. To show the function behavior, Fig. 2.15 is plotted for Eq. 2.8 with r=s=1, T_n=3600s, Y_{hc}=0.7 and Y_{lc}=0.5. The message number has a minimum at N=6. It means that for these parameters with N=6 the message number will be at the lowest possible level. It indicates that the sample should be taken at every $T_s = T_c / N \approx 508s$. The sample number in which the message number becomes minimal is called optimal sample number. As an example for r=3 and s=7, N corresponding to the minimum g is equal to 8.

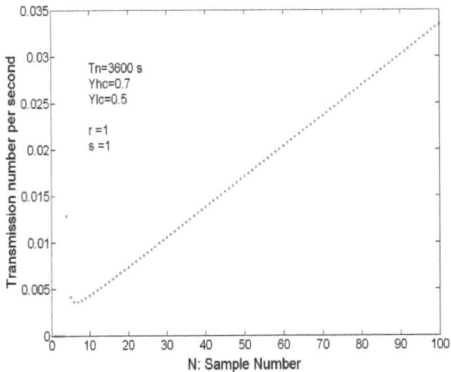

Figure 2.15: Number of message transmission corresponding to each sample number in central structure of Fig. 2.14.

The following example shows how this method works. In Fig. 2.14 it is supposed that the s=1, now we start to increment "s" one unit at each step and find the corresponding optimal sample number. The result is presented in table 2.1. Now suppose that when "s" increases from 1 to 10 the sample number remains equal to 6. In this case the message number will be equal to g(6,1,10) ≈ 0.188 which in comparison with the message number at s=1 (g(6,1,1) ≈ 0.0036), the message number raised up about 4.91 times. But the message number at s=10 with the optimal sample number calculated in Table 2.1 raises up just 3.55 times (g(12,1,10) / g(6,1,1) ≈ 3.55) which is about 28 percent less than message number without optimal sample number. On other words the dynamic of the method is so that it tries to compensate the message number increase partly by adding the sample number and consequently reducing the message number.

Now "s" is taken equal to one and fixed then the same procedure is done by increasing "r". The results are presented in Table 2.2. To decrease the message number the sample number is decreasing which match conceptually with the expectation. Because as it is already mentioned the sensor sends message at each sample time, therefore to reduce the message number from sensor to center the sample number should be declined. But in this case the message cannot be less than specific value because it is already confined with conditions in Eq. 2.5. Comparison of these two assessments shows that it better to choose the center close to the sensor rather than to the actuator.

Table 2.1: Optimal sample number corresponding to hops number from center to actuator

s	Minimum g	N
1	0.0036	6
2	0.005	8
3	0.0062	8
4	0.0073	9
5	0.0084	10
6	0.0094	10
7	0.0104	11
8	0.0113	11
9	0.0123	12
10	0.0132	12

Table 2.2: Optimal sample number corresponding to hops number from sensor to center

r	Minimum g	N
1	0.0036	6
2	0.0056	6
3	0.0074	5
4	0.0091	5
5	0.0107	5
6	0.0123	5
7	0.014	5
8	0.0156	5
9	0.0173	5
10	0.0189	5

2.2.7) Sample number selection in AWSAN

In AWSAN, the communication path is modeled in Fig 2.16. The sensor measures the control parameter and compares it with the limits whenever measured variable goes over the limits, the sensor sends a message to the actuator. The message number in time interval of T is equal to $(T/T_d) \times 2 \times r = (T \times (p(N)+1)/T_c) \times 2 \times r$ and in time unit it is expressed as function "h" in Eq. 2.9. From another point of view, Eq. 2.9 can be computed from Eq. 2.8 by putting r=0 as well. To explain this point of view we can suppose that the center in the central network is fragmented to the small processors

2. Development of AWSAN

and they are merged with the nodes to shape the autonomous network. In central network the center sends message to the actuator the same way the sensor does in autonomous network. This resembles that the center and sensor are merged, therefore "r" in Fig 2.14 disappears and its value in Eq. 2.8 becomes zero, on other words it means (h(N,r)=g(N,0,r)).

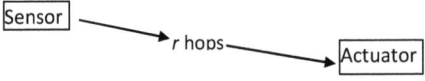

Figure 2.16: Autonomous network structure

$$h(N,r) = (p(N)+1) \times 2 \times r \times f_c$$ Equation 2.9

As it is explained in section (2.2.4), increasing the sample number decreases the message number in AWSAN. The Eq. 2.9 explains this concept mathematically as well. When the sample number increments from N=i to N=i+1 happens, the actuator frequency drift p(N) declines (Fig 2.12) and in follow the message number (h(N,r)) decreases in Eq 2.9. The change of message number for one unit increment in sample number is formulated in Eq 2.10. For this reason increasing the sample number causes the reduction of transmission energy on one hand and on other hand it causes increase of process energy consumption for taking more samples. If the reduced transmission energy is looked as "saved" energy and increased process energy consumption is considered as "cost" energy, the optimal sample number is where the saved energy is still more than cost energy or in other words when saved energy minus cost energy is minimum but positive. At this point if the sample number increases one more step, the cost energy will be more than what it will be saved; therefore the sample number will not be optimal any more. In this way the total network energy consumption will be reduced because by increasing the sample number, the transmission energy decreases versus increasing of process energy and since the transmission energy is much bigger than process one the summation of transmission and process energy will decreases.

To compare the transmission energy and process energy, they should be expressed by a common measurement unit. For simplification and avoidance of calculation of both cost and saved energy in different types of nodes and networks, it is assumed that the average of process energy is the energy measurement unit. For further computation it is assumed that transmission energy is "e" times of the process energy. In example computation the "e" is taken equal to 10 which for the nodes we worked with is reasonable value.

In Eq. 2.10, Δh represents the message number change. When Δh decreases, in addition to reduction of transmission energy and increment of process energy in sensor node, the process energy in intermediate nodes decreases because they will forward fewer messages. This reduction is equal to (Δh×(r-1)/r). By consideration of this factor the total saved energy will be equal to the summation of the reduced transmission energy and retransmission energy in intermediate nodes which can be formulated as (Δh×e+Δh×(r-1)/r).

$$\Delta h_i^{i+1} = h(i,r) - h(i+1,r) =$$
$$(p(i) - p(i+1)) \times 2 \times r \times f_c = \Delta p_i^{i+1} \times 2 \times r \times f_c$$

Equation 2.10

The cost energy is equal to the consumed process energy for taking the sample. In time unit, the sample number is equal to the sample frequency (f_s) and regarding to Eq. 2.3 the sample frequency is equal to N×f_c. When sample number increments from i to i+1, the sample number increase will be equal to ((i+1)-i)×f_c=f_c. Regarding to the assumption that the relative energy measurement unit is the average process energy consumption, the cost energy is equal to f_c as well. It is already mentioned that the condition for the optimal sample number is the maximum N where the saved energy is still larger than the cost energy. This inequality is expressed in Eq. 2.11.

$$\Delta h \times e + \Delta h / r \times (r-1) > (i+1-(i)) \times f_c \Rightarrow$$
$$\Delta p_i^{i+1} \times ((e+1) \times r - 1) \times f_c \times 2 - f_c > 0$$

Equation 2.11

As example for this method the former parameters and system model are applied to calculate the numerical value of sample number. With Y_{hc}=0.7, Y_{lc}=0.5 and T_n =3600s, the sample number are calculated for different number of hops and the results are presented in Table 2.3. For example when r=2 then the sample number is equal to 15 (N=15) and T_s=T_c/N ≈ 190s would be the optimum sample period.

Table 2.3: Maximum N values for which inequality of 12 is valid for different "r".

r	1	2	3	4	5	6	7	8	9	10
N	12	15	19	21	23	25	27	28	30	31

This optimum sample number can still be changed regarding to other criteria as well. For example after finding the sample number if its actuator frequency drift is not still acceptable, the sample number can be increased but it should be known that the system is paying for such improvement by energy consumption. For example in Table 2.2 when r is 2, N=6 but with respect to Eq. 2.6 and Fig. 2.12 the actuator frequency increases about 150%. If this oscillation is not acceptable, N can be increased to 8 and actuator frequency drift reduces to about 80%.

2.3) Wireless nodes

For physical implementation a tiny wireless node which is designed to be applied as a wireless sensor node is considered. Particularly "Tmote Sky" from Moteiv Company is taken as sample. They have the option that to be provided with integrated temperature, Humidity and light sensor or without them. This mote offers some expansion terminal which could be used for integration with other type of sensors or be used as processor unit of the actuators.

To evaluate the autonomous structure functionality, the WSAN is simulated in next chapter. By the simulation the autonomous structure functionality is compared with central one. For simulation the technical specification of the nodes and simulator should be determined. In fact, offering a model for nodes is not just step for entering physical implementation but also it is prelude for the simulation in next chapter.

2.3.1) Hardware specifications

"Tmote Sky" in Fig 2.17 is essentially made of a microcontroller, radio chip, antenna, memory and the communication interfaces. The node is supplied with two batteries size "AA" [38sky]. Briefly the components and Tmote Sky specifications are listed in below [38sky]:

- Microcontroller:
    - Microcontroller MSP430 f1611 (8MHz, 10k RAM, 48k Flash)
    - 8 channels Analog to Digital Convertor (ADC)
    - 2 channels Digital to Analog Convertor (DAC)
    - Direct Memory Access (DMA) controller
    - Supply Voltage Supervisor
    - Event driven type with sleep and wake up mode
    - 6µs transition from the sleep mode to the wake up mode
    - 2 built-in 16 bits timers

- Wireless transceiver: CC2420, Chipcon, 250kbps data transfer rate over 2.4GHz compliant with IEEE 802.15.4 (PAN-LR)
- Antenna: miniaturized on board with 50m indoors and 125m outdoors range
- Ultra low power consumption
- USB interface
- 16-pin expansion connection

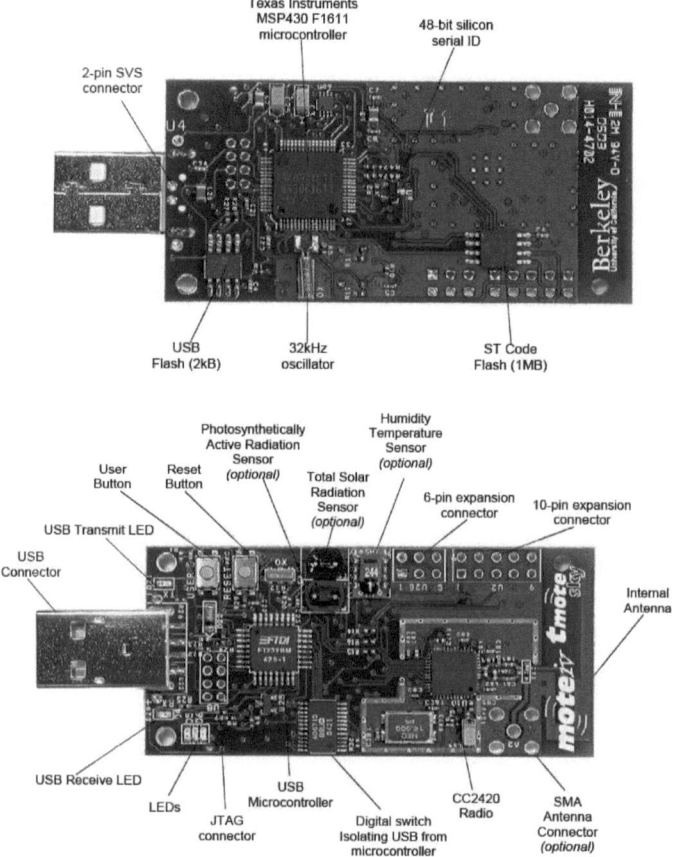

Figure 2.17: Tmote Sky components [38sky]

2. Development of AWSAN

Inside the microcontroller an oscillator module is integrated which is externally supplied by a 32 KHz crystal oscillator. This module includes Digitally Controlled Oscillator (DCO) and a high frequency crystal oscillator [39msp]. This module provides the base for the low cost and low power consumption. This block provides the Auxiliary Clock (ACLK), Main clock (MCLK) for CPU and Sub Main Clock (SMCLK) for peripheral modules [39msp]. By help of these clocks one active mode and five low power modes are realized. Normally the microcontroller is in sleep mode, when a hardware or software interrupt, called as event, happens the clock source will be turned on and stabilized by DCO in 6 μs. Then the microcontroller goes to the awake mode, performs the interrupt service routine as its task and then goes back to the sleep mode [40msp]. In this way the microcontroller consumes low energy and thereby it is known as ultra low power consumption and event driven microcontroller. Another feature of "MSP430 f1611" which supports the low energy consumption is DMA. By DMA the data can be moved from one memory location or peripheral to another one without the central processor involvement [39msp].

The microcontroller includes 8 channels of 12 bits ADC, two channels of 12 bits DAC, I^2C bus, two USARTs, and 48I/Os pins [39msp]. Some of these peripherals are wired to expansion connectors on Tmote Sky board illustrated in Fig 2.18. The expansion connectors provide the possibility to integrate this board with other electronic devices such as sensors or actuators. For example suppose that a pneumatic equipped window receives a message to close. The microcontroller activates a digital output on expansion connector which is connected to the pneumatic board to close the window. Another example could be integrating of a sensor with the nodes. The sensor value can be read via ADC channel on expansion connectors and processed by the node microcontroller and transmitted by radio chip on the board.

2.3.2) Software specification

The introduced wireless nodes are categorized as embedded system. The embedded system does not work just by hardware but also it needs software to bring all parts together to function. TinyOS is the tiny operation system for these wireless nodes. TinyOS is a real-time and an open source operation system. Its core is very small in volume so that it fits within the tiny nodes hardware limitations. TinyOS scheduler is based on the First Input First Output (FIFO) model but the priorities of the scheduled items are different. The hardware interrupt are called as "event" with higher priority and the deferred computation program are called as "Task". The tasks can be preempted by the events but not by other tasks. Since TinyOS is real-time operation system, all the tasks are run within a limited interval of time [36Gay]. The microcontroller is normally in the sleep mode, when it receives a software or hardware interrupts it executes the assigned program and then goes back to the sleep mode [37tiny].

NesC is the programming language in TinyOS environment. This language is based on the C programming language and it is component oriented language. Component oriented programming makes the programming less complicated and makes the diagnosis easier. An application is made of Configuration and Module components. The components are wired to each other through interfaces. Hardware initialization for functioning is done by calling their designated components [37tiny].

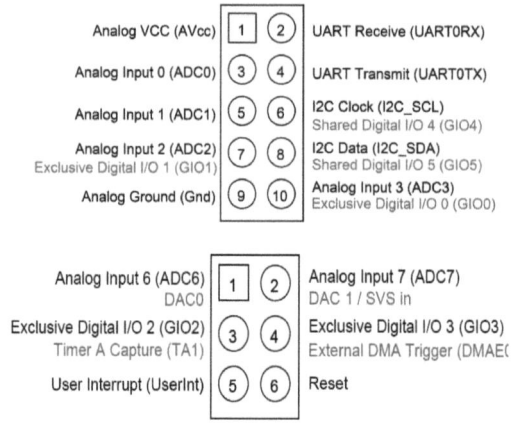

Figure 2.18: Expansion connectors on Tmote Sky [38sky]

Chapter 3

Simulation and Comparison

Abstract:
In this chapter the autonomous and central WSAN are simulated in order to compare their functionality in terms of the network energy consumption with its distribution over the nodes and the networks robustness. In the first section, Prowler is introduced as simulator. It is explained how this simulator works and which general modifications are applied in order to use this simulator for this project. For the network energy consumption evaluation a task of communication between two nodes as sensor and actuator is defined. Fifty random topologies are simulated with communication of two hundred pair of nodes in each. The second section finishes with the evaluation of the simulation results for the central and autonomous network. For robustness comparison, in the third section an application for watering of an apple orchard is defined. The humidity dynamic of this orchard is modeled and implemented in Prowler. By implementing noise over communication channel, the system output reaction and deviation is considered for conclusion. At the end of sections two and three, discussion parts are opened to consider the further aspects of the simulation results especially in logistic systems. In fourth section, the scalability of the networks is compared regarding to the network structure impact on the message number increase. In the last section the main element of the network which causes the different behavior of the network structures is revealed. It is explained that under which circumstances this element will change. By this, it is shown that the advantages of the network structures are relative. A specific point is extracted in which the networks merits become even.

3.1) Prowler

3.1.1) Introduction

In this chapter WSAN is simulated to evaluate and compare the functionality of autonomous and central network structure. The probabilistic wireless network simulator (Prowler) is chosen as simulator [44prow]. Prowler can simulate nondeterministic nature of wireless communication channel by applying normal distribution function. Obviously by ignoring this function it can be used for deterministic applications as well e.g. testing ideal situation or reliable results. In prowler all nodes can be identified by their ID label and theoretically it can incorporate unlimited nodes with a dedicated application for each node, but in TOSSIM as another simulator it is not possible to define dedicated applications for the nodes [42toss][43sidh]. It means that all the nodes should have the same application which does not make sense for the automation process applications.

Prowler is MATLAB base and event-driven like TinyOS. It includes Graphical User Interface (GUI) for visualization. Prowler program body is basically formed by three modules: Prowler, Radio and Application. These modules have interactions with each other via functions which are called events and commands. Application module incorporates events: "Init_Application", "Packet_Sent", "Packet_Received", "Collided_ Packet_Received" and "Clock_Tick". Radio module puts the MAC layer and radio propagation model into operation. This module includes the events "Init_Radio", "Channel_Request", "Channel_Idle_Check", "Packet_Receive_Start", "Packet_Receive_ End", "Packet_Transmit_Start" and "Packet_Transmit_End". The main module which is called prowler is activated by the commands such as "StartSimulation" and "InsertEvents2Q". By application module the possibility is provided to define the designated tasks for different nodes and for WSAN.

All the events and commands are stored and scheduled by event handler in event queue. The schedule algorithm is based on the Earliest Deadline First (EDF) strategy. By running Prowler, firstly an initialization process starts in which the nodes and network topology are identified and the radio and application parameter are read from Sim_Param.m file and assigned. Then by Set_Clock command during initialization the Clock_Tick event is inserted in the events queue. In this event designated application can be defined for each node, for example sending message.

This Simulator is developed based on the Berkeley MICA mote. This mote is essentially made from an 8-bit, 4 MHZ Atmel ATMEGA 103 microcontroller, 4KB RAM,

3. Simulation and Comparison

128 KB program memory and a radio chip. The Radio chip of this mote is RFM TR1000 with 40kbit/s transmission rate over 916.5 MHz carrier frequency [42toss]. The simulator time step is taken equal to the transmission time for a bit of data which is equal to 1/40000 second. All the timing for scheduling the tasks or simulating MAC layer are expressed based on this time step.

The MICA mote implements a simple carrier sense multiple access protocol for MAC layer. For transmission a packet it waits for a random time interval between a minimum and maximum value, and then it checks the transceiver channel. If the channel is free, it transmits the message otherwise it waits for a random back-off time interval and then checks the channel again. If it is not occupied, it transmits the message. The Waiting time is assigned in "Channel_Request" function. In "Channel_Idle_Check", it is checked that the channel is idle or not by checking the radio status of the node. If the node status is already set in transmit or receive mode, then the channel will be realized as busy and waits for back-off time. The maximum and minimum for waiting and back-off time are defined in sim-param.m file. The routing algorithm is integrated in "Channel_Idle_Check" event as well. If channel is idle, the "Packet_Transmit_Start" event will be put in events queue for the sender node and "Packet_Receive_Start" will be set for all nodes which are supposed to receive the message including the next intermediate node for forwarding the message.

In "Packet_Transmit_Start" the node status will be set into "transmit" mode and the "Packet_Transmit_End" event is scheduled for this node based on the required transmission time depending on message length and radio bandwidth. When this event is run the node status changes to "idle" and "packet_sent" as application event will be set. In the receiver node when the "Packet_Receive_Start" is called, if the receiver node status is "idle", then it will change to the "receive" and "Packet_Receive_End" event will be set for time after reception time which is equal to transmission time. If the receiver node status was not "idle", collision is happened and the message will be dropped.

Prowler has the capability to detect data collision in two ways. The simpler model used in this work, checks the status of the receiver node. If the receiver node is in "receive" mode, the data collision is realized. Prowler uses a channel model for computation of received signal strength. The default channel attenuation function is set to be $1/(1+x^2)$. In this function the parameter x is the distance from the sender node antenna. At distance x from sender node the signal power is calculated by multiplying the transmission power by the attenuation function value. This is in ideal reception power. When the noise over communication channel is taken into account, the random noise generator value is subtracted from this ideal reception power.

The default routing algorithm in Prowler is simple flooding which is implemented in Radio module. In follow it is explained how the SCAR is implemented instead of flooding. After finishing the simulation MATLAB retains the global event list and prowler provides a "sysstat.m" file to analyze the happened events and to extract the nodes and the system statistics such as number of sent, received and collided messages.

3.1.2) Modification

Three types of modification are done in Prowler in order to use Prowler for this project. The first and basic one refers to changing the platform from MICA to Tmote Sky. The second modification is for implementing SCAR as routing algorithm and the third main modification is about incorporating the wireless node energy model for the energy consumption evaluation. Some other extra elements are added to Prowler for different applications which are explained in the application related section.

With changing the platform the most significant modify is related to the new radio chip. As it is explained in the previous chapter, CC2420 chip broadcasts data with 250 kbit/s by 2.4 GHz. Regarding to this the time unit is changed to 1/250000 second. Correspondently wait time and back-off time are refined. The running time of the simulation is restricted to the specific period of time e.g. ten days.

SCAR routing algorithm is implemented in the Radio module but before that some preparations are done. The first preparation is implementation of the topology in the topology file by calling a "Topology_Generator" function. By this function, the topology graph is produced then regarding to the assumed channel attenuation function in Prowler, the Channel Attenuation Values (CAV) are assigned as the weight of the topology graph edges. Finally the SCAR coordinates are assigned corresponding to the nodes ID. In the message header two extra fields are appended which contain the next forwarding node coordinates and the destination node coordinates. The SCAR algorithm is applied in Channel_Idle_Check.

To evaluate the node and the network energy consumption, an energy consumption model of the radio chip physical layer is implemented. This model includes the reception and transmission energy. A consumption energy field is appended to the node status field in which the node energy consumption for sending or receiving messages are accumulated therefore this field shows the total node energy consumption. By sysmstat.m, the nodes energy consumption are extracted from the consumed energy field in node status and it is used for further analyzes, for example computing the network energy consumption or its distribution over the nodes.

3. Simulation and Comparison

To calculate the reception energy by radio chip the usual formula of power multiple by time is used in Eq. 3.1. The supply voltage (V_{DD}) and the reception current (I_{rec}) are extracted for radio chip from CC2420 data sheet [45cc24]. The time for sending message is computed by the message length and radio bit rate. In this case the TinyOS packet length is equal to 40 bytes (320 bits) and the inverse of the radio chip bit rate is 1/250Kbits/s. The reception power calculation is presented in Eq. 3.1.

$$P_{rec} = V_{DD} \times I_{rec} \times t =$$
$$2.1 \times 0.0197 \times 40 \times 8 \times (250000)^{-1} = \qquad \text{Equation 3.1}$$
$$52.95 \ \mu J / packet$$

Table 3.1: Energy consumption for each bit transmission in 8 levels

Level	I_{trans} (mA)	W_{trans}/per bit (μJ/bit)
1	8.5	0.0714
2	9.9	0.083
3	11.2	0.094
4	12.5	0.105
5	13.9	0.117
6	15.2	0.127
7	16.5	0.138
8	17.4	0.146

Another part of node energy consumption model is for the transmission. On the contrary to the MICA radio chip, CC2420 supports the multi level transmission power. The transmission energy is divided to 8 levels. When the receiver node is located nearby, the sender can send a message by a lower transmission power. This multi level output is implemented in transmission energy model. Each level has its current consumption (I_{trans}) which is shown in Table 3.1 [45cc24]. Concerning to the radio chip power supply, the transmission energy per bit for each level is computed and presented in Table 3.1. To realize that in which transmission power level the node is broadcasting, the ideal reception power is calculated based on the distance of receiver from sender by the attenuation function. Based on this computed energy the required transmission power level is realized.

It should be reminded that in the energy model just the radio chip energy consumption is taken into the consideration and the microcontroller energy consumption called as process energy consumption is not computed because it is normally much smaller than the transmission energy so that it could be ignored. The wireless nodes usually are in the sleep mode. By an interrupt occurrence, they wake up and perform the assigned task to that interrupt and then go back to the sleep mode. Moreover, this transition from sleep mode to awake mode takes about 6 μs, which does not cause considerable energy consumption. Therefore this energy consumption for transient between two modes is not present in energy model as well.

One of other changes is adding a "Packet_Lost" event to the radio chip module. This event is separated from collision although during the collision the packet will be lost too, however it is not the only source of the message lost. The message lost can happen when the noise over channel becomes so strong that the reception power becomes lower than Reception_Limit. In this case the message is sent and the receiver is capable to receive message, but message is lost during transmission. All possible message lost events are marked by "Packet_Lost" event and saved for further analyzes.

3.2) Energy consumption

Energy consumption is one of the comparison and evaluation items between the autonomous and central network. The energy consumption is evaluated with two points of view. First one is the network energy consumption in general and the second point of view is the network energy consumption distribution over the nodes. Since in the central network every message should pass through the network center, it is expected that the energy consumption in the central network becomes more than autonomous network energy consumption. This prediction is based on this argument that if a sensor-actuator direct path does not pass directly through the center, then the message has to diverge from direct path to pass through the center which causes more energy consumption for message forwarding. On the other hand, passing all messages through the center causes that the nodes around the center become involve in message transmission in each message transmission between sensors and actuators, therefore these nodes consume more energy than the nodes which are further from the center.

3. Simulation and Comparison

This means that the total network energy consumption is not evenly distributed over the nodes. By conducting the energy consumption simulation this two phenomena are verified [48eng][51eng].

3.2.1) Simulation

In order to simulate the WSAN for the energy consumption evaluation, some preparations should be done in advance. The Simulator, Prowler, explained in the former section is taken into the service. As preparations, the network topology should be generated and implemented in the simulator and then a scenario for the simulation should be defined.

Topology generation

To make network topology, two options are considerable: Deterministic and Random topology. Deterministic topology means that it is already known where the nodes are located or the nodes are arranged in the particular way. Random topology is generated by scattering the nodes randomly and then based on their connectivity the network graph is extracted. In this way the generated random topology is not directly representative of any predetermined network but it could represent a network topology. The reason for objecting with the deterministic topology is that the final result will be dependent on the applied particular topology and the result follows special mind pattern of the nodes distribution. Therefore the result cannot be generalized or in other words it cannot be considered as an approximated estimation of general result. But the random topology could symbolize any network topology and it represents that there is no intention to drive a particular result. In order to become independent from the topology and generalize the result, the simulation is repeated for number of random topologies and their averages are extracted. In this simulation 50 random topologies are generated and the simulation is run for them.

To generate the topology a 1×1 plane in MATLAB is taken as base. Although in reality the physical world is three dimensional, however the two dimensional coordinates is sufficient for calculation because the transmission energy consumption is a function of distance. I this way the required distances between nodes are modeled without involving with the complexity of another dimension existence. By the random generator function in MATLAB a set of 400 random numbers are generated. The 200 of them are used for the horizontal, "x", coordinate of the nodes and the rests are taken as vertical, "y", coordinate. In this way 200 points are randomly scattered in 1×1 plane. Fig. 3.1 shows one example.

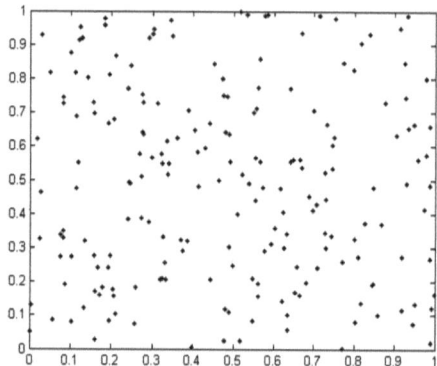

Figure 3.1: 200 random points scattering example

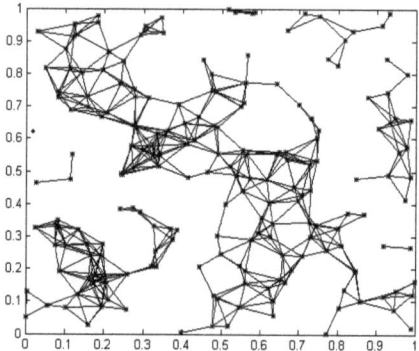

Figure 3.2: connected points with distance less or equal 0.1

After this phase it is supposed that the normalized maximum radio coverage of the nodes in this plane is equal to 0.1. It means that the nodes with distances equal or less than 0.1 are connected. By connecting these nodes a graph is shaped which can be realized by an adjacency matrix [26gary]. Fig. 3.2 shows such graph. This graph is not necessarily a "connected graph". Based on the connected graph definition there is at least one path from each node to another one on the connected graph [28jung]. It means that the generated graph may have different separated components. In reality one character of the network graph is connectivity because if some nodes are isolated, they cannot be called as a network and they have to be considered as another network. To attain the topology for simulation, the component with larger number of points is

3. Simulation and Comparison

separated and the other components are eliminated. This procedure can be done by Deep First Search (DFS) algorithm [27foul]. The result of such process for graph in Fig. 3.2 is illustrated in Fig. 3.3. In Each topology the remained nodes number will be different and it is not necessarily equal to the same number of scattered nodes (200). In order to produce the 50 random topologies for simulation repetition, this procedure is repeated for each random topology generation.

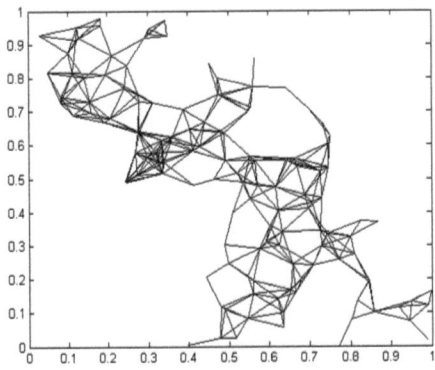

Figure 3.3: connected points with distance less or equal 0.1

Simulation scenario

In order to simulate the central and autonomous network for the energy comparison, a scenario is planned. In this scenario each sensor and corresponding actuator forms a pair. In the autonomous network the nodes in each pair send message to each other directly and in central network they communicate with each other through the center. Suppose that there is a network with "n" nodes and each node could be a sensor or actuator, to make a pair of nodes there are "n-1" options for a node correspondent. In this way the total pair number is equal to n×(n-1)/2. In the case that the network nodes number is large, the process for taking all the pairs into the computation will be too long. To avoid such computation the sample mean is used. In this way, limited numbers of pairs are randomly chosen as instances, and then their average is calculated. This average will be approximation of the average of the all pairs but since in comparison of the central and autonomous network the ratio of the averages is important, the distance of approximated average from the real one will be less effective in the final result. Practically in this simulation 200 nodes are randomly chosen as a sample group of the sensors (Group S) and 200 nodes as a sample group of actuators (Group A).

Simulation common basis

In an autonomous network each sensor from the group S sends a message to its corresponding node in the group A and the energy consumption for this transmission is computed. In the central network the node in group S sends a message to the center and the center sends message to the corresponding node in group A and the energy consumption is calculated and stored. In this process the center location has impact on the energy consumption. If during simulation the common base is not applied for two network structures, the result will be biased with outcome of the imposed situation and it can put the validity of the result under question. To compare the functionality of two networks they should be in their best condition as a common base. Regardless of this simulation, in the central network the best place for the center location is where the summation of the distances from other nodes to the center becomes minimal [26gary]. This node is called "median" of the network and it is found by "Dijkstra" algorithm [26gary].

For forwarding a message from one node to another one, by different routing algorithm different path could be chosen and consequently the energy consumption will be different. In order to have common base, a common routing algorithm must be applied. In this simulation the SCAR algorithm introduced in the second chapter is implemented in Prowler and used for both network structures.

In energy simulation the multi level transmission power output of the radio chip is implemented. To achieve this purpose the ideal reception power between nodes by the channel attenuation function is computed based on the distances of the nodes. The computed ideal reception power determines which level of transmission power is suitable for the transmission.

If data collision happens in middle of the path, the message will be lost and the rest of the path will not be taken for the energy consumption computation. To avoid this kind of phenomenon which causes miscalculation none two node pairs send message simultaneously, the pairs send message after finishing the previous transmission.

3.2.2) Conclusion

Network energy consumption

The network energy consumption is equal to the accumulation of all transmission energy consumptions between the selected pairs of nodes; therefore the transmission energy consumption between the nodes of each pair is calculated and stored.

3. Simulation and Comparison

Corresponding 200 node pairs, there are 200 computation steps which are presented on the horizontal axis of Fig. 3.4 and Fig. 3.5. As it is already mentioned in order to be independent from the network topology, 50 random topologies are simulated. Accordingly, the transmission energy consumption at each step on the horizontal axis is the average of the transmission energy consumption of 50 node pairs of 50 different topologies, each one from one topology. The network energy consumption as the accumulation of the transmission energy consumption is depicted in mill joule (mJ) on the vertical axes of Fig. 3.4.

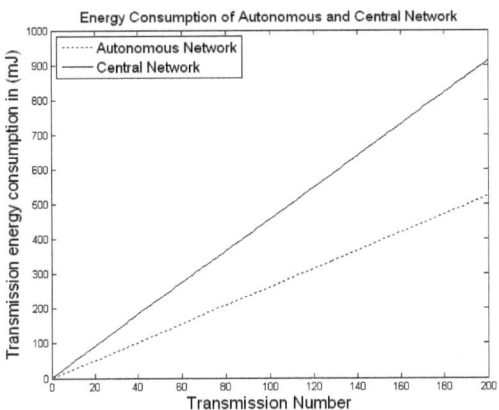

Figure 3.4: Network energy consumption for 200 random pair of nodes in 50 random topologies

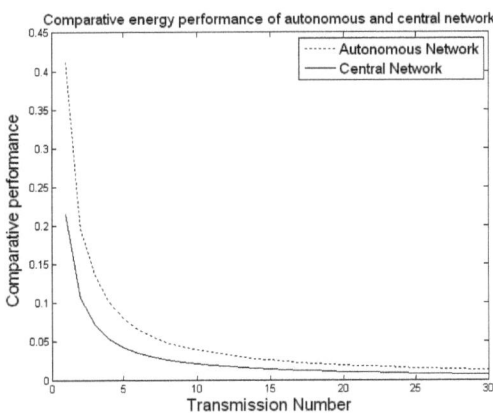

Figure 3.5: Comparative energy performance of autonomous and central network

This figure shows that the central network for communications between sensors and actuators consumes more energy than the autonomous network. At the end of the simulation, the total energy consumption ratio of the central to the autonomous network is equal to 1.74 which means that the central network is consumed about 74 percent more than the autonomous one. In other words the central network for performing an assigned task consumes more energy than the autonomous network, therefore if the energy supply of the nodes is limited like a battery, the sustainability of the central network will be less than autonomous network. On the other hand, consuming more energy by central network implies that the central network performance in comparison with the autonomous one is less. Moreover this shows that the performance is proportional to inverse of network energy consumption. The comparative performance is the ratio of autonomous network performance to the central one which correspondently is proportional to the central network energy consumption to the autonomous one. The performances graph versus the message number increasing is depicted in Fig. 3.5. In order to have better resolution, the horizontal axis is truncated to the smaller transmission number. As it can be seen in this figure the autonomous network offers better performance in comparison the central network.

Network energy consumption distribution

After finishing the simulation, the total energy consumption in both networks is available. The question is how this network energy consumption is distributed over the nodes. The importance of this question is related to the sustainability of the network. To answer this question one topology is chosen randomly as instance and its node energy consumptions with both network structures are sorted on the horizontal axes. Sorting algorithm operates so that the node with the highest energy consumption locates in the middle of the horizontal axes and the nodes with the minimum energy consumption places at the corners. The sorted result is illustrated in Fig. 3.6. From this figure it can be observed that the node energy consumption in the central networks varies between about zero and twenty mJ but in the autonomous network the node energy consumption varies between about zero and six mJ. This observation implies that the network energy consumption is more evenly distributed over the nodes in the autonomous network rather than in the central network. After identifying the nodes with the high energy consumption in the central network, it can be seen that these nodes are located around the center. From this observation it can be inferred that the nodes around the center have more jobs to do than the others, therefore they go out of the resources and services more likely and faster.

3. Simulation and Comparison

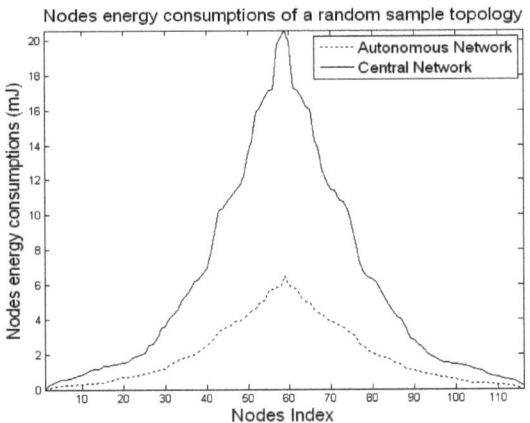

Figure 3.6: Nodes energy consumption of a sample random topology

Table 3.2: Average of the node energy consumption and their standard deviations of 50 random topologies

	Central Network	Autonomous Network
Aveage of nodes energy consumption (mJ)	5.02	1.44
Average of standard deviations	5.17	1.4

To investigate this result in general term, the average of nodes energy consumption for 50 topologies are computed. Then the standard deviation of the nodes energy consumption in each topology is calculated and finally their average for 50 topologies are computed and presented in Table 3.2. By comparison of the average of the nodes energy consumption it is inferred that the nodes in the central network consume about 3.5 times more energy in average rather than the nodes in the autonomous network. The standard deviation of the nodes energy consumption shows that in the central network, the network energy consumption is distributed over the nodes with larger deviation. It approves that some nodes in the central network consumes much larger energy than others and in the case that the node power supply is limited e.g. supplied by batteries, the energy supply will be depleted much faster in some nodes than the others in the central network. This phenomenon decreases the maintenance interval and increases the cost ad efforts for the network maintenance.

3.2.3) Discussion

Process energy

In the autonomous network the process energy for computation of a control task is provided by the nodes' battery while in a central network the process energy is provided by the center which usually has unlimited power supply. In Table 2 the difference between the nodes average energy consumption is (5.02-1.44=) 3.58 mJ. It implies that if up to 3.58 mJ is devoted to the computation in each node, the autonomous network still consumes less energy than the central network and is recommendable structure.

Time domain

In above simulation the time element is not taken into the consideration and it is assumed that both networks have the same number of the message transmission. Regarding to the sampling section in second chapter, the message number in the central and autonomous network is not equal in period of time. In the central network the message number per time unit is proportional to the sample frequency plus actuator frequency whereas in the autonomous network the message number in time unit is proportional to the actuator frequency. Regarding to this, the energy consumption of each pair of nodes per time unit in central network is proportional to the F_s+F_c and in the autonomous network is factor of F_c for each pair. The network energy consumption for all the pairs will be summation of all the nodes energy consumption. Since F_s+F_c is greater than F_c then the aforementioned difference in energy consumption between two network in time period will be greater. In other words the computed energy consumption for the autonomous network is related to longer time of operation in comparison to the central network. The central network consumed the computed energy in shorter time.

Regarding to the argument in above the nodes depletion will be faster in terms of times, which means less sustainability for the central network. It implies that less sustainability of the central network is not only the network energy is not evenly distributed over the nodes but also it originates of faster energy consumption by the network.

Logistic system

Another discussion is whether this result can be generalized to other autonomous or central organization structures with the different type of elements e.g. logistic system? If the nodes are taken similar to the entities like warehouses, retailers, ports and so on which offer services and the communication is taken to be as transport system between

3. Simulation and Comparison

them, the above simulation can resembles of autonomous logistic system and central one. Accordingly the energy consumption resembles the transport system fuel consumption. With this point of view it can be claimed that transportation in the central network costs more energy in comparison to the central network.

The network energy consumption distribution result showed that the nodes closer to the center are busier than the further nodes. With above resemblance it implies that the entities around center have to deliver more services than the entities are further from the center. If we define throughput of the entities as ratio of offered services to demanded services, this phenomenon can lead to reduction of throughput. Subsequently the bottleneck problem appears around the center while in the autonomous system regarding to more even distribution of the task over the system these problems are less likely to happen.

3.3) Robustness comparison by an orchard example

One of the factors for comparing the autonomous network with the central one is the robustness. The question is how resistance these networks are in the environment with a significant probability of communication failure. In this section it is evaluated which network structure for WSAN is more robust or less sensitive to the noises and can resist more against the noise [50robs]. The expectation is that the central network will be affected stronger by noise over communication channel than autonomous one. Especially when the sensor node is far from the center its communication number will be larger therefore it would be more likely to be influenced by noise. In this sense losing communication between the nodes should be started from the further nodes and it gradually progresses to the nodes closer to the center by increasing the noise strength. On the contrary it is expected that the autonomous network shows better resistance against communication lost which can happen by noise. If these claims could be confirmed, they would provide strong evidence for advantages of autonomous network to the central network.

In this section a case study is conducted to compare the robustness of the AWSAN versus CWSAN. It is assumed that 100 apple trees in a garden example are shown in Fig. 3.7a. The trees are distributed on a circle indicated in Fig. 3.7b. There are five humidity sensors and five water valves. Each sensor is correspondent with one valve. The humidity of the orchard should be controlled by opening and closing the water valves. The humidity sensors are implemented on Tmote Sky nodes which are explained in the former chapter. The water valves are equipped with Tmote Sky. For the rest of trees one node is devoted to each two trees as a connection bridge for establishing connection between sensors and actuators. It is assumed that the distance between nodes is so that each node is just in a connection with two closest nodes to itself and not further.

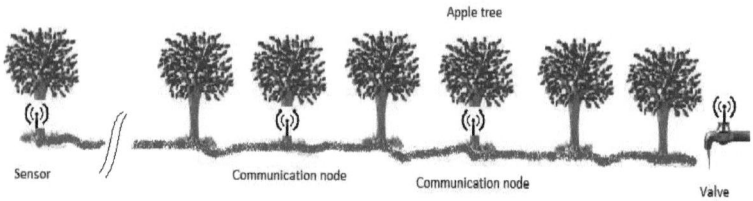

Figure 3.7a: Apple trees with valves, communication nodes and sensor

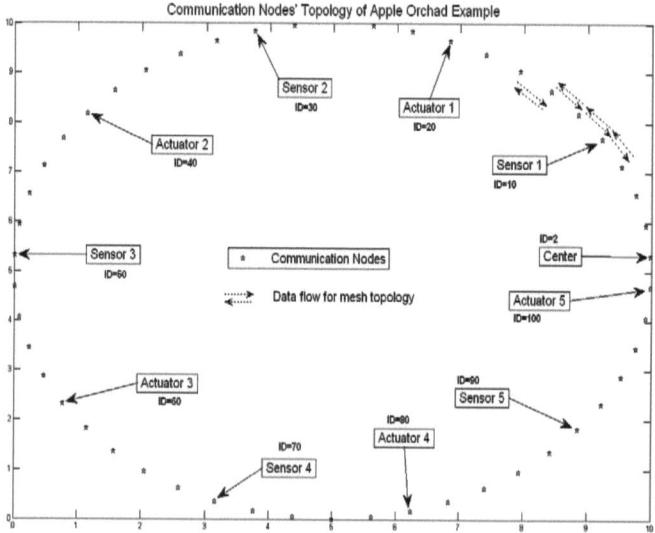

Figure 3.7b: Wireless nodes topology in the experimental apple orchard

3.3.1) Simulation preparations

In order to simulate this orchard with Prowler some preparations should be made. The first one is providing the system dynamic model to simulate the sensor values while the valves are open or close. In follow the humidity relation between trees is expressed and

3. Simulation and Comparison

the state space model is derived from this relation and implemented inside Prowler. With this model the sensor values are taken as humidity of the trees on which the sensors are attached and the orchard humidity is defined as average of these five sensors. The orchard humidity is called system output too. Another preparation is the noise model. The noise will be applied on the communication channel with different levels of strength. Afterward a simulation scenario is needed to be the guideline for conducting the simulation. In this simulation because of the simplicity of the orchard topology, it is avoided to use the SCAR routing algorithm introduced in former chapter. As a substitute a simple routing algorithm which forwards the message to the next immediate neighbor in left or right is used.

Orchard humidity state space model

Suppose that $\phi_i(k \times T)$ is a state variable representing the humidity of the ith tree at time $k \times T$. It is assumed that the humidity of the ith tree at time $(k+1) \times T$ is stated in Eq. 3.2. In this equation α is humidity permeability coefficient. It represents the humidity of the tree i^{th} which permeates to neighborhood trees which are indicated by index of $(i+1)$ and $(i-1)$. The part $(-2\times\alpha)\times\varphi_{(i)}(k)$ in $(1-2\times\alpha)\times\varphi_{(i)}(k)$ represents the humidity reduction of ith tree which is permeated equally into its immediate neighbors.

$$\varphi_i(k+1) = \alpha \times \varphi_{(i-1)}(k) + (1-2\times\alpha) \times \varphi_{(i)}(k) + \alpha \times \varphi_{(i+1)}(k) \quad \text{Equation 3.2}$$

Since the humidity reduction cannot be more than the tree humidity therefore alpha should be equal or smaller than 0.5 ($\alpha \leq 0.5$). In this simulation α is chosen to be 0.4. Likewise the parts $\alpha \times \varphi_{(i\pm1)}(k)$ represent the humidity absorbed from the $(i+1)^{th}$ and $(i-1)^{th}$ trees by i^{th} tree. The process represented by Eq. 3.2 is type of Markov process [46ogat] because each state depends just on the state variables in one time step back, not more.

Since there are 100 trees, a state vector presented in Eq. 3.3 with 100 state variables is taken. The index variable indexes represent the tree number on the clockwise. The general state space model is presented in Eq. 3.4. Matrix A is the transition state matrix of the system. Matrix A is derived from Eq. 3.2 and presented in Eq. 3.5. It can be seen that in Fig. 3.7b, the 100th tree and the first tree are neighbors and the humidity will be permeate between these two trees too, therefore for covering this boundary condition, the elements A(1,100) and A(100,1) become equal to α in the matrix A.

$$\varphi_{100\times1}(\kappa) = [\varphi_1(\kappa), \varphi_2(\kappa), \ldots, \varphi_{99}(\kappa), \varphi_{100}(\kappa)]' \quad \text{Equation 3.3}$$

$$\begin{cases} \underline{\varphi}(k+1) = A \times \underline{\varphi}(k) + B \times u - F \\ \underline{y}(k+1) = C \times \underline{\varphi}(k) \end{cases}$$

Equation 3.4

$$A_{100 \times 100} = \begin{bmatrix} 1-2\times\alpha & \alpha & 0 & 0 & \cdots & 0 & \alpha \\ \alpha & 1-2\times\alpha & \alpha & 0 & 0 & \cdots & 0 \\ 0 & \alpha & 1-2\times\alpha & \alpha & 0 & \cdots & 0 \\ \vdots & \ddots & \ddots & \ddots & \ddots & \ddots & \vdots \\ 0 & \cdots & 0 & \alpha & 1-2\times\alpha & \alpha & 0 \\ 0 & \cdots & 0 & 0 & \alpha & 1-2\times\alpha & \alpha \\ \alpha & 0 & \cdots & 0 & 0 & \alpha & 1-2\times\alpha \end{bmatrix}$$

Equation 3.5

In Eq. 3.4, "u" represents the input of the system and it is normalized to be equal to 1. Matrix B in Eq. 3.4 is the system input matrix which presented in Eq. 3.6. Matrix B has 100 elements. Generally elements are equal to zero except the 20th, 40th, 60th, 80th and 100th elements. On the trees with the same number the valves are attached, these elements are correspondent to the water valve status on these trees. If any of valves is open its corresponding element in matrix B is equal to 1 otherwise it is set as zero. For example if valve (i) (ID=i) is open, B$_i$=1 and when it is closed, B$_i$=0. During simulation in order to open or close a valve, its corresponding element in Matrix B is changed between zero and one.

$$B_{100 \times 1} = [B_1 \cdots B_{100}]', \forall \ i \in \{20, 40, 60, 80, 100\}$$
$$if \quad Valve(i) \ is \ "on" \ then \ B_i = 1 \ otherwise \ B_i = 0$$

Equation 3.6

$$F_{100 \times 1} = [F_1 \cdots F_{30}, F_{31} \cdots F_{70}, F_{71} \cdots F_{100}]'$$
$$F_1 \cdots F_{30} = 1/30; F_{31} \cdots F_{70} = 1/40; F_{71} \cdots F_{100} = 1/35;$$

Equation 3.7

$$\underline{y}_{5 \times 1}(k) = [\varphi_{10}(k), \varphi_{30}(k), \varphi_{50}(k), \varphi_{70}(k), \varphi_{90}(k)]'$$

$$C_{5 \times 100} = \begin{bmatrix} C_{1,1} & \cdots & C_{1,100} \\ \vdots & \ddots & \vdots \\ C_{5,1} & \cdots & C_{5,100} \end{bmatrix}$$

Equation 3.8

$$C_{1,10} \ \& \ C_{2,30} \ \& \ C_{3,50} \ \& \ C_{4,70} \ \& \ C_{5,90} = 1 \quad and \quad \forall i,j \ \ C_{i,j} = 0$$

3. Simulation and Comparison

This system has some loss which occurs by some factors for example evaporation. This humidity loss is modeled by matrix F in Eq. 3.4. It is assumed that the elements of matrix F are different because of the geographical diversity, for example in one area there is more sunlight or the soil quality is different. These different ranges of values are presented in Eq. 3.7. As it is stated and Fig. 3.7b shows, the sensors are located on the trees 10,30,50,70 and 90, likewise the state variable of these trees are taken as the system output. These state variables are extracted by matrix C, output matrix, in Eq. 3.4. The values of matrix C elements are presented in Eq. 3.8.

This state space system model is implemented in Prowler and from the state variables, the sensor values are driven. Equation 3.2 is made by transformation to the discrete domain from the continuous domain by taking sample at each T seconds and this sampling period is different from the sensor sapling period. If we suppose that the sensors take samples at k×T time, the above model in simulator is called at each k×T by event handler then the new state vector and the system output is computed. More description will be offered in the simulation scenario explanation.

Noise model

In unit plane in Fig 3.7b, the shortest distances between the nodes are fixed and are equal to 0.628. Regarding to this shortest distance between two nodes and the channel attenuation function implemented in prowler, the ideal reception power will be equal to 0.717 (P_{rec_ideal}=0.717) with taking P_{out}=1. In order to have a range for reception power variation the "RECEPTION_LIMIT" is set to be 0.65. In other words it means that if the reception power becomes less than "RECEPTION_LIMIT", the message will be rejected. With this "RECEPTION_LIMIT", only the left and right direct neighbors are able to receive a message.

Now with the P_{rec_ideal} the real reception power after the noise implementation can be computed. The Eq. 3.9 shows the formula for P_{rec} calculation.

$$P_{rec} = P_{rec\_ideal} - P_{rec\_ideal} \times (Random\_Noise)$$
$$Random\_noise \longrightarrow \sigma \times randn(n)$$

Equation 3.9

In Eq. 3.9 the random noise is generated with *"randn"* which is normal distribution function in MATLAB with average 0 (μ=0) and standard deviation of 1 (STD=1) (σ_n=1). The general form of the normal distribution function is depicted in Fig. 3.8. Since this figure resembles a bell it is called bell curve. As it can be observed with μ=0 & σ_n=1, 68.26 percent of generated numbers are between $\mu \pm \sigma$. The value of σ indicates how wide or narrow the bell waist is. When the generated random set is multiplied in a coefficient, the average of data set remains constant but its standard deviation differs. By larger σ the bell waist will be wider it means 68.26 percent of generated number are located in larger intervals and with smaller σ the waist will be narrower and the generated number will be closer to the mean value.

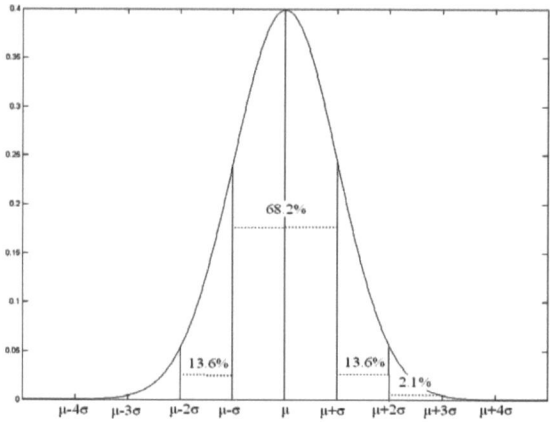

Figure 3.8: Bell curve

The random generator multiplied by σ represents the ratio of the noise power to the ideal power reception. By multiplying the P_{rec_ideal} the noise power is calculated. Since the real power reception cannot be larger than the ideal power reception, in Eq. 3.9 the noise power is reduced from the ideal transmission power. In fact by this formulation the interval [μ μ+σ] from bell curve is used. After this calculation if the P_{rec} becomes less than "RECEPTION_LIMIT", it means that the message is lost. In order to have more or less powerful noise, σ is changed.

Simulation scenario

As it is mentioned, Eq. 3.3 is the system model in discrete domain. This model is made by sampling from a continuous model. This sampling is different from the sampling period by sensor. If we assume that the sampling period of the system is one minute, the sensor sample period regarding to the pervious chapter is determined to be 20 minutes. To implement the sensor sampling period, it should be expressed by simulator time unit. In Prowler periodical events are scheduled for the sensor to compute the state variables. Whenever the events are called by scheduler of Prowler, the state variables are computed from the last computed irritation to the present one. For example if the preceding system states variable computation is performed at 54 minutes of the simulation time and the next one is called at 70 minutes, with the sample period equal to one minute, then the state variable should be calculated for 16 irritations.

After the state variables are computed and the sensors value is determined, in the central network, the sensors send a message containing their values to the center. The Center compares the sensor values with the predetermined limits boundary. If they are beyond the limits band, the center sends a message to the corresponding actuator to change their status. This is done by changing the matrix B elements. In above topology in Fig. 3.7b actuator(i) is corresponded with sensor(i). With the autonomous structure the communication scenario is different. After determining the sensor values, the values will be compared with the limits boundary incorporated in the sensor program, if they are beyond the limits, the sensors send message to their actuators directly. Actuators after receiving the message from their correspondent sensors, change their status by changing the B_i values in matrix B. If the communication in any step is lost the system keep on with the same status it already had. For example if the message which supposed to close the valve status is lost, the valve remains open and the humidity increases.

Another important aspect is about the applied random set. If one random data set for noise is taken for the simulation, the conclusion will be dependent on the certain random data set. To avoid such dependency, the simulation is repeated for 10 random sets and the final result is driven based on the average of the 10 round simulations for each AWSAN and CWSAN. Each round time period is set up for 10 days. In AWSN the noise standard deviation increases from 0 to 1 in 50 steps for each round. During the simulation we noticed that CWSN is very sensitive to the noise, therefore these 50 steps are done between $\sigma=0$ and $\sigma=0.25$. It means that we took increment step 0.005 in CWSAN in comparison to 0.02 in AWSAN. In each round the sensors output are stored in a Microsoft Excel file for further process in the next subsection.

3.3.2)   Conclusion

By observation and analysis of the simulation results two conclusions are drawn. The first shows that the impact of noise over communication channel is distributed over autonomous network more evenly than over central network structure. The second is that the autonomous network is more robust against the noise over the communication channel. In below these results are presented, formulated and explained.

Noise impact distribution

Comparing the sensor outputs in different levels of noise discerns a pattern for the network behavior versus the noise in the autonomous and central network structure. Figures 3.9 and 3.10 depict the sensor 1 and sensor 3 outputs. It can be seen when σ is increased from 0 to 0.02, the sensor output does not change noticeably. Since the differences between outputs are not distinguishable, the outputs are cut off to show their existence and similar behavior. By boosting the noise level to σ=0.06, the sensors 1 and 3 react differently which is illustrated in Fig. 11 and Fig.12. Comparing Fig. 3.11 and Fig. 3.12 shows when the noise is boosted, in the central network the sensor 3 output oscillates in a larger interval than the output of the sensor 1 with the same level of noise but in the same situation the sensor 1 and sensor 3 outputs do not fluctuate and do not deviate from each other very much in the autonomous network. These observable facts happen for two reasons, one is for message lost between the sensor and center and another one is the message lost between the actuators and center. For example when the center sends an instruction to the actuator to close it, but the message is lost in between the actuator remains open, the humidity rises up and the sensor measures larger values or when the sensor sends the measured value to the center but the message is lost in between, then the center cannot realize that the humidity is over the limit band consequently it does not send any instruction to the actuator. All these impacts appear on what the system output or more precisely on what the sensors measure.

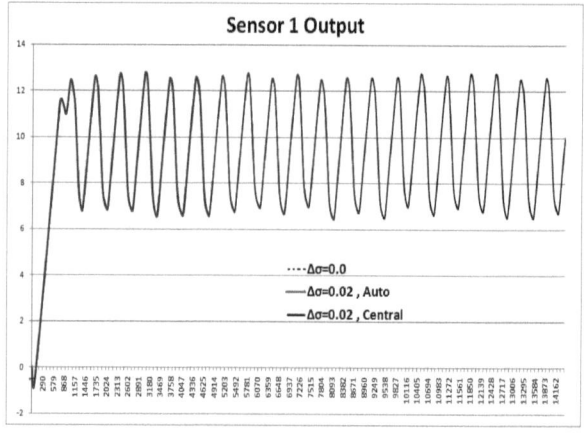

Figure 3.9: Sensor 1 output for σ=0 & σ=0.02 (differences are not distinguishable)

3. Simulation and Comparison

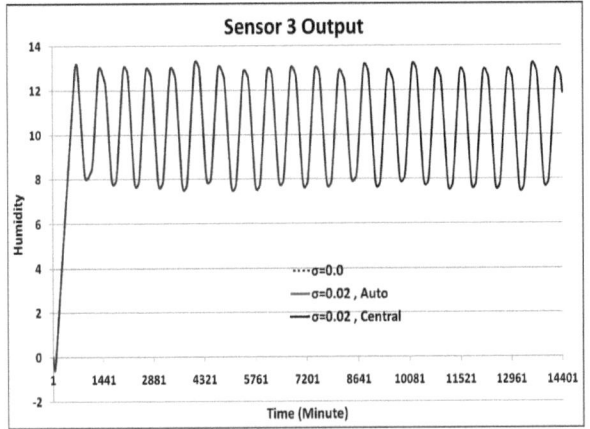

Figure 3.10: Sensor 4 output for σ=0 & σ=0.02 (differences are not distinguishable)

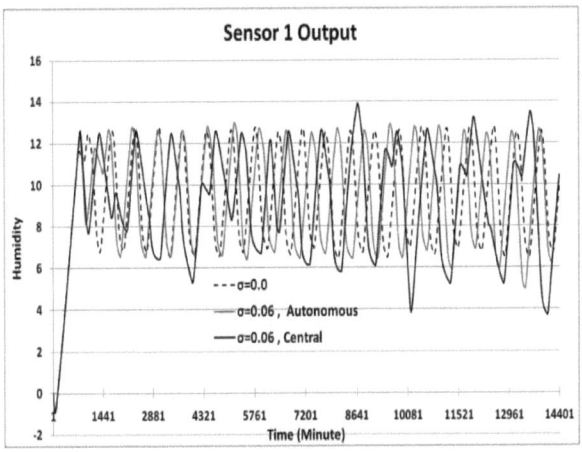

Figure 3.11: Sensor 1 output for σ=0 & σ=0.06

As it can be seen in Fig. 3.7b, the sensor 3 is located further from center than the sensor 1. It implies that a part of system which is further to the center is affected by noise stronger than the parts close to the center while in the autonomous network difference between the noise effects cannot be observed. In other words the noise impacts over communication channels emerge more evenly in the autonomous network comparison to the central network. Conceptually the distant parts of system need more number of transmissions to communicate with the center than the closer parts; therefore under the same noise affect probability their communications are more influenced.

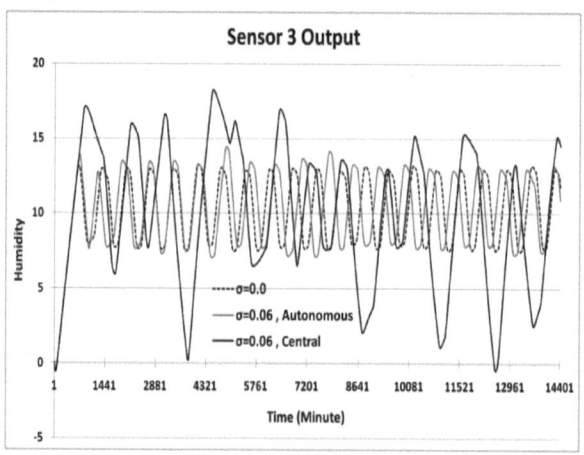

Figure 3.12: Sensor 3 output for σ=0 & σ=0.06

This observation is important in terms of reliability and scalability. When the environment situation changes this result shows that it is more likely to lose the connectivity and the control of distant parts. This means the network structures are not the same reliable as each other. On other hand from this observation it can be concluded that the central network in harsh environment is restricted for scaling. In the first chapter it is mentioned that the central network encounters scalability problem because it may go out of peripherals or resources. However, the scalability in this section is in terms of adding nodes in distant from center which could represent the geographical scalability. From this point of view the center can still provide peripherals and resources for scaling the network, but environmental condition like noise presence does allow it. In other words, the central system under risky situation loses its integrity faster and the system decomposition starts from distant parts.

The sensor 1 output in Fig 3.13 and the sensor 3 output in Fig 3.14 approves the same aforementioned pattern in both networks. With σ=0.12 in the central network the sensor 3 output is out of the limit range and accordingly out of the control but the sensor 1 output still shows oscillations. Although the oscillation is out of the limit range, however it infers that some messages can be exchanged between the sensor 1 and the

3. Simulation and Comparison

center or the center and the actuator 1. This difference in appeared behavior of the sensor1 with σ=0.12 means that more messages in the noisy environment of closer nodes to the center has chance to survive in the central network I comparison to the further nodes. The sensor 1 and the sensor 3 output in the autonomous network are not very different, although in the Fig. 3.14 the oscillation range sounds smaller but it is because of larger scale of vertical axes.

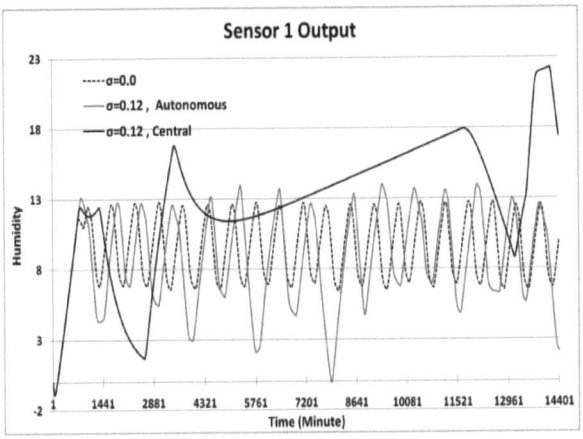

Figure 3.13: Sensor 1 output for σ=0 & σ=0.12

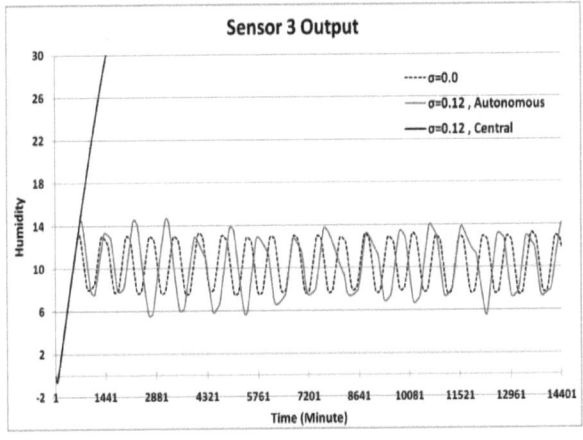

Figure 3.14: Sensor 3 output for σ=0 & σ=0.12

Sensitivity and Robustness

In this part the humidity of the orchard is taken into consideration as the system output. There are five sensors; the humidity of the orchard at each moment is defined as average of these sensors measurements. When the noise level is zero, the system output is taken as reference for further analysis. The humidity set point is assumed to be 10. The Fig 3.15 shows the system output which slightly fluctuate around the assumed set point. It can be seen that the small amount of noise (σ=0.02) does not alter the system output and the humidity in both networks is almost the same as when the noise is not applied.

In this section the impact of noise on the system is formulated by Sensitivity and Robustness. Sensitivity refers to system reaction against the system surrounding condition changes. When a surrounding condition alters, the system with stronger reaction is called more sensitive than others. If the changes in surrounding conditions are unpleasant and leads to the divergence from proper functionality of the system, the sensitivity does not convey positive concept, so the robustness is used as inverse of sensitivity. For example if noise rises over communication channels, the system which shows weaker reaction is less sensitive and more robust.

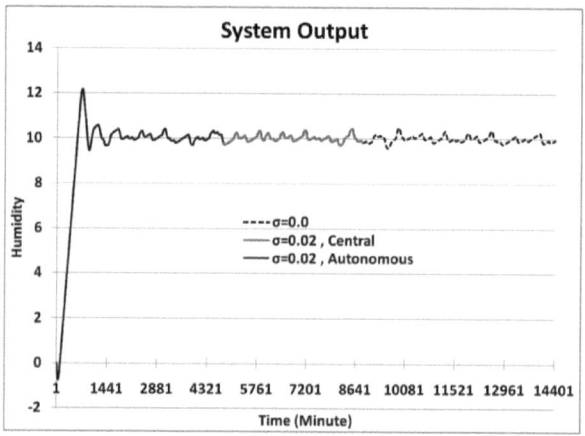

Figure 3.15: Sensor 1 & 3 and system output for σ=0 & σ=0.02

Now we look at the system output with higher noise level in Fig. 3.16. By σ=0.06 the orchard humidity with the central automation structure swings in wider band than the autonomous one. By doubling the noise level, Fig. 3.17 indicates that the control over

3. Simulation and Comparison

the humidity in the central network is totally lost. In the autonomous network the output swings just in wider range and it is still around the system output without noise. This observation implies that the central network is more sensitive and less robust to the noise than central network.

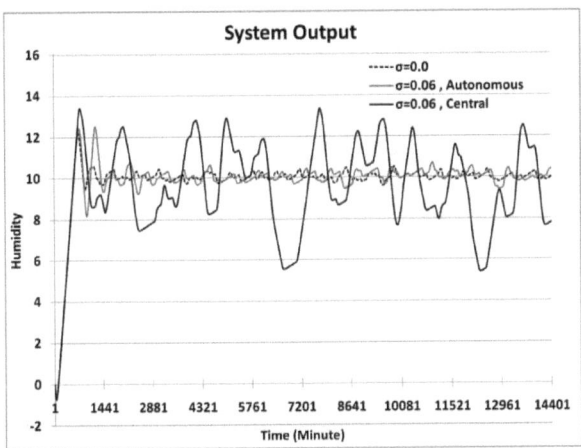

Figure 3.16: Sensor 1 & 3 and system output for σ=0 & σ=0.06

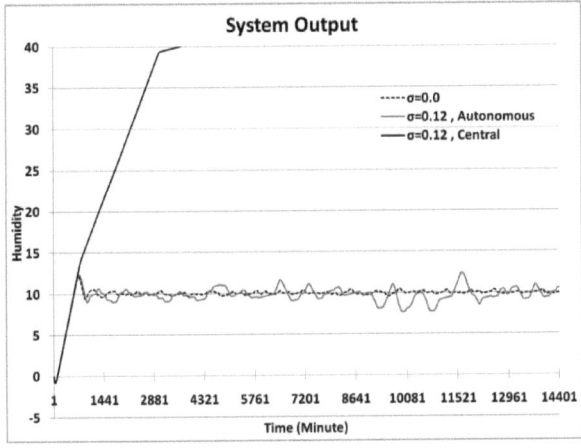

Figure 3.17: Sensor 1 & 3 and system output for σ=0 & σ=0.12

Nevertheless this system output is just taken from one round simulation with a specific random data set for the noise simulation therefore the result is highly dependent on this data set. To generalize the result, such kind of dependency should be avoided. To achieve this goal, the simulation is repeated 10 times with different random sets. The system output average for a single level of noise is computed after the first time the output passed the lower limit value. With the 10 rounds repetition of the simulation with 10 different random sets for each level of noise, there are 10 system output averages. Now by taking the average of these 10 averages, the general system output for each level of noise is calculated and this computed system output is less dependent on the random data set and delegates the system behavior more. In above paragraph, according to the one random data set it is stated that with a stronger noise power, the central network is less robust in contrast to the autonomous one.

Now to generalize the claim and turn it into conclusion, the system output error is computed and depicted in Fig. 3.18. The error is defined as subtraction between the general system output at each level of noise and the system output without noise. These errors are sketched in Fig. 3.16 for both network structures. It can be seen that when the noise become stronger, the absolute value of the error in the central network becomes larger than the error in the autonomous network. The error of averages for the autonomous network remains around the zero whereas in the central network the error trends to infinity for the smaller levels of noise. The error behavior of these two networks indicates the high sensitivity and low robustness of the central network.

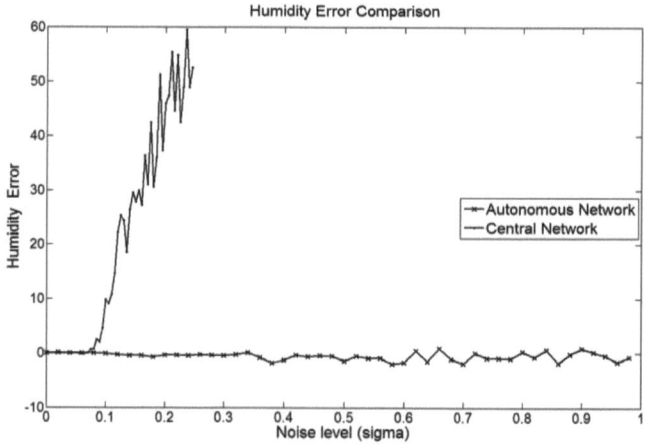

Figure 3.18: Humidity error of averages in the central ad autonomous network

3. Simulation and Comparison

There is another factor which should be considered in order to have more precise conclusion. Figure 3.19 shows the system output for autonomous network with higher level of noises. The point is that the outputs swing in wider ranges but they do not necessarily trend toward infinity like the outputs in the central network. Mathematically it is possible to have two signals with the same period and the same average but with two different fluctuation bands. Comparing these two signals by their averages does not lead to anything, because both of them will have the same average. One of the means for comparing these two signals and showing their amplitude deviation from their average by digits is their standard deviation calculation.

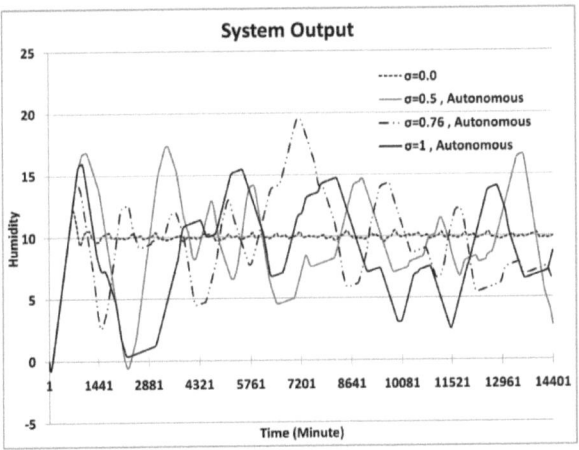

Figure 3.19: System output in autonomous network for different levels of noise

The comparison between the system outputs by standard deviation is illustrated in Fig. 3.20. For each round of simulation, at each level of noise the standard deviation of the system output is computed. Then the average for ten rounds simulation is taken. The final result for each level of noise is presented in Fig. 3.20. This figure confirms the previous conclusion in above paragraph in terms of the standard deviation as well. When the noise power increases, in central network the output diverges from set point larger than autonomous network. This result confirms again that the autonomous network is more robust than the central network versus noises which represents the probability of communication failure.

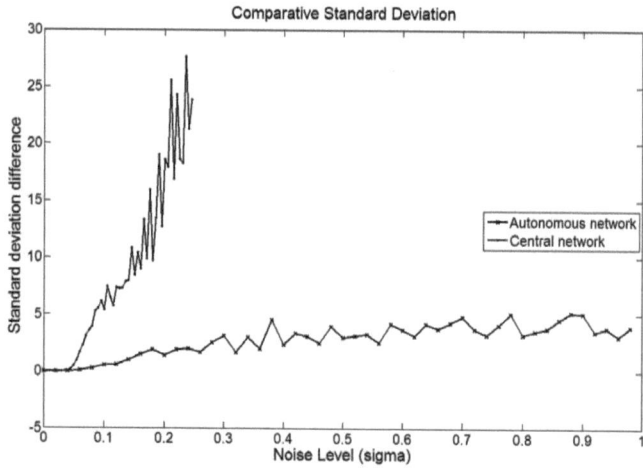

Figure 3.20: Autonomous and central system output standard deviation

3.3.3) Discussion

By the simulation results assessment, two conclusions are drawn. The first conclusion is that the impacts of the environmental factors which cause more likely communication failure are distributed over the autonomous network more evenly than over the central network. The second conclusion is that the autonomous network is more robust against the same environmental factors.

Similar to the discussion in section 3.2.3 for the logistic system, the above networks and simulation results can be used to draw the results for the central and autonomous logistic system in general. The sensors, which provide information for the decision making, resemble the suppliers and the actuators which water the trees can be seen as retailers which deliver the goods to customers. The intermediate nodes could be the warehouses, ports and so on. Again similar to section 3.2.3, the communication can be seen as transportation in the logistic systems. With such resemblance, the simulation conclusions imply that if there is a high probability of transportation failure or risk in a logistic system, the autonomous structure for the system can be more immune against it and achieve better service delivery. On the contrary under a risky environment, a system with central structure reacts to the environmental conditions stronger and faster in negative sense. Its service as a logistic system output reduces. In section 3.5, it will be explained that such difference between the autonomous and central network is rooted

3. Simulation and Comparison

in number of communication or transportation they need to fulfill a task. The communication number is not constant in different situations. Sometimes the autonomous system has less number of communication and sometimes the central system and above application is so that the autonomous has less number of communications, therefore is influenced less by noises or interruptions.

3.4) Scalability

In the first chapter it is mentioned that one of the distributed system features is scalability which refers to the scale of the system in terms of size, geographical distance and administratively. Size of the system is defined as number of subsystems. In Center-Periphery model presented in Fig 1.1 adding subsystems is limited to the number of peripherals provided by the center. But in the hierarchical structure like Fig 1.2 it is possible to dispose this limitation and add another layer. Although the center can still face with the lack of resources, however the peripherals limitations do not appear in the hierarchical structure of a central system. In this section by assuming to have enough resources the capability of two networks for scaling is evaluated. Regarding to the optimum sample number section in the development chapter, it is showed that the autonomous network has more capability for scaling. It is concluded that in scaling AWSAN, the message number increase can be decreased by adding the sample number while this option in the central network does not have the same effect. Two case studies is conducted for the autonomous and central network according to the hypothetic application in section 2.2 and the outcome is evaluated.

Central network

For the central network the communication path model is shown in Fig. 2.14. The system scales with increasing the hops number between the center and the sensor or actuator. As case study, the example system and its transfer function in section 2.2 for one pair of sensor and actuator is taken into attention again. The scalability test starts by holding the distance between the center and sensor fixed (r=1) and increasing the distance between the center and actuator indicated by s from 1 to 10. For each hops number (or s), the message number versus sample number is sketched in Fig. 3.21. The same test is done with holding s=1, fixed distance between the center and actuator, and increasing r from 1 to 10. The computed message number of this test is illustrated in Fig. 3.22. In both figure to have clear figure, r and s are sketched just from 1 to 5.

Comparison of these two figures shows that the message number increases with larger step by increment of the distance between the sensor and center (r) in comparison to the distance increase between the center and actuator (s). Because the base in finding the optimum sample number is the minimum message number, therefore the method provides a sample number for the greater hops number so that the message number becomes minimal. In other words if the system with the new and higher hops number works with the non-optimal sample number the message number will be more than when the new sample number is extracted by the method.

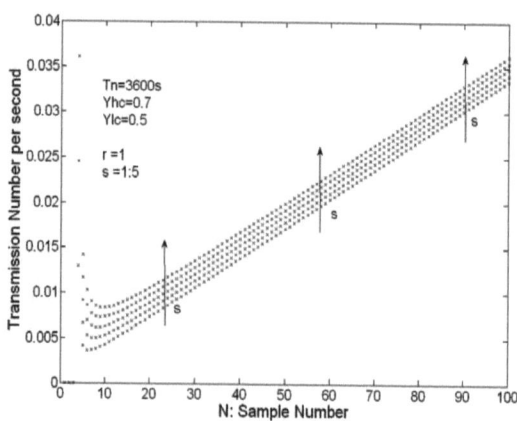

Figure 3.21: The message number versus optimal sample number by distancing the actuator and center for the example system

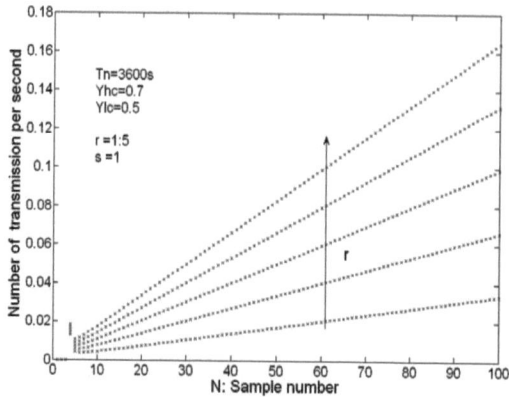

Figure3.22: The message number versus optimal sample number by distancing the sensor and center for the example system

3. Simulation and Comparison

Table 2.1 shows that when distance between the sensor and center (r) increases, the optimal sample number decreases. This reduction can be explained so that in the central network the sensor sends message at each sample period and the message will be repeated r times. Then message number will increase by higher r. To reduce the increased message number, the optimal sample number method offers lower sample number. Likewise, table 2.2 indicates that when the distance between the center and actuator (s) increases, the optimal sample number increases. This observation can be justified with following explanation. Increasing s raises the message number in the center-actuator wing. Since the center sends the message just at the crossing point of the system output with the limits and its frequency is equal with double of actuator frequency, therefore to reduce the message number the actuator frequency should be reduced. Actuator frequency reduction means smaller actuator frequency drift. It is already discussed in the second chapter that to reduce actuator frequency drift, the sample number should be increased to decrease the actuator frequency drift.

The difference between the center-sensor and center-actuator wings for scalability appears here. To reduce the message number, for the center-sensor wing sample number should be decreased and for another wing the sample number should be increased. Increasing sample number is more feasible than decreasing the sample number because sample number is limited from bottom. To verify this claim Eq. 3.10 is drawn from Table 2.1 and Table 2.2. The left side shows the message number increase ratio when the r increases to 10 with the optimal sample number and the right side is the same value for increasing s. It can be seen that the message number increase in the sensor-center wing is 50 percent more than the message number increase when the distance between the center and actuator increases.

$$[g(5,10,1)/g(6,1,1)] > [g(12,1,10)/g(6,1,1)] \cong 5.25 > 3.66 \quad \text{Equation 3.10}$$

From the above explanation it is concluded that if there is any option t choose the center location, it is more efficient to choose the center to the sensor than to the actuator. This consideration resembles <Autonomous structure for the network which is explained in the next paragraph. In other words this conclusion is stating that the nodes which have more communication traffic should be closer to each other than the others.

Autonomous network

From another point of view, the autonomous network is a distributed central network so that the center is decomposed to small parts and each part is located inside a node. This point of view links the conclusion in above paragraph with the autonomous structure. In other words the sensor and center become so close that the center-sensor wing is disappeared (r=0). Consequently the message transmission between sensor and center is eliminated which leads to two advantages: Firstly the message number related to the center-sensor wing is omitted from the total message number and secondly the sample number can be increased without concerning about decreasing of the message number in another wing. In fact by an autonomous network the aforementioned difference between the sensor-center and actuator-center wings is totally resolved in side of the actuator-center wing.

From the energy consumption point of view, in the autonomous network by increasing the sample number the process energy consumption increases but in the central network the transmission energy consumption increases, which is much greater. On one hands it means that for the specific amount of energy consumption the sample number can be larger in the autonomous network. With higher sample number, lower actuator frequency and better control quality can be achieved on another hand it means that the central network has to consume more energy for achieving the same control quality as autonomous network. Overall, scaling the network size in the autonomous network can be dealt by increasing the sample number in order to decrease the network traffic but in central one there is no such option.

3.5) Autonomous or Central

In the first chapter, the autonomous structure is conceptually explained and defined. The subject is followed in the second chapter by explaining of this structure development for the wireless sensor actuator network. In this chapter WSAN is simulated with the central and autonomous structure and then their performance and functionality are compared. The conclusions depict the network with autonomous structure works relatively more efficient than the network with the central structure. Since there is no absoluteness in the scientific thinking way and each solution is dependent on the surrounding conditions, a question arises here, is the autonomous network always more efficient than central one? If no, under which condition are the privileges of the autonomous network justified? In this section it is tried to answer these questions for WSAN.

Message number

To answer these questions, the former sections of this chapter are reviewed to discern the fundamental element of the network which leads to difference appearance between these two structures. In the energy consumption evaluation it is concluded that with the

3. Simulation and Comparison

autonomous structure the network energy consumption is less. The measured energy consumption is base on the transmission energy consumption. The network energy consumption is proportional to the message number directly. If there is more message number in a network, the network energy consumption will be more and vice versa. Therefore it can be concluded that the message transmission number is less in the autonomous network than in the central network.

On the other hand, in the robustness evaluation section it is shown that if the central and autonomous network works under noisy environment, the autonomous network resists more against the noise interference. The noise distribution probability is the same for both networks, but in one network more transmission messages are influenced. This infers that the more affected network includes greater number of the messages. This claim can be easily verified by considering the probability calculation formula. The probability is equal to the ratio of happened events to possible events. When the probability of two occurrences is the same but the happened events for one of them are more than another one, mathematically it is concluded that the possible events is greater too.

In the scalability section it is discussed that scaling the central network with the hierarchical structure is more limited than autonomous network, however scaling in the central network causes more message number because the sample number cannot be decreased as strong as it can be increased in the autonomous network to compensate the message number increase.

From above three paragraphs it is inferred that the message number has a key role in functionality and properties of the network. A network with less message number consumes less energy, is more capable of scalability and more robust with less influence by noise. Regarding to this fundamental element, message number, which has impact on the behavior of the network, it should be evaluated that under which condition the message number will change. When the message number changes in a network, the aforementioned appeared conclusions will change too. In follow it is explained that under which situations the message number varies from application to application in the autonomous network.

Conditional decision making

In the first chapter it is mentioned that sometimes to make decision in autonomous network the entities needs to have information from other entities as well. As example, suppose that there are windows and heater in a room and the temperature of the room should be controlled. When room gets cold the heater turns on to increase the temperature. The temperature sensor sends the measured parameter to the heater to inform it that the room is cold. The heater should check the window whether it is close or open, because if the window is open, turning on the heater does not help and it wastes energy. Because of this it sends a message to the window and request to receive the window status from it. In such cases the decision making process becomes conditional and it is dependent on some criteria i.e. information from other entities.

In the central WSAN, since all sensors send their data to the center and center knows the last status of the actuators, the criteria of having information from other nodes does not lead to send and receive more messages. However these criteria just increase the data processing effort in the center and demands for more resources. Accordingly the message number remains constant in the central network when the criteria number increases. But in the autonomous network the decision maker should send a request message to the nodes to receive the required information. If an entity decision making process includes two criteria it means that it should ask from two other entities about their statuses. This phenomenon causes the increase of the message number in the autonomous network. As it is mentioned already, this variation of the message number has impact on the functionality and properties of the autonomous network.

In energy consumption evaluation and robustness evaluation, it is assumed that the nodes decision making is unconditional. It means that the actuators make decision just based on the received information from their corresponding sensors. Now if we suppose there are some criteria for decision making by the actuators, the message number in the autonomous network will increase and at one point it will become greater than the message number in the central network. After this point, the autonomous network with greater message number consumes more energy, it will be affected by noise more and its scalability can face with more problems as well.

Hop number

However, there is still one point about the criterion number as a basic parameter for further computation and comparison. The criterion number cannot be used for computing the message number because the criterion number cannot be associated exactly with a specific message number. For example if we assume that a decision maker depends on one criterion from a node in five hops away, the data acquisition leads to the 10 message transmissions (multiplied by two because of round trip). If the same node is two hops away, the same criterion causes four messages transmission. It means that for one criterion, different number of messages can be required. For this reason, to have a common and a unique base to compare different criteria and compute the message number, the hop number is taken as a basic parameter. To clarify the above concepts, the below computation with an example in WSAN is discussed. The message umber versus the hop number is calculated and the crossing point of the autonomous network over central network is calculated.

3.5.1) Message number in autonomous and central model

Now in the autonomous WSAN we suppose that a sensor-actuator communication path is modeled in Fig. 3.23. The system is Linear Time Invariant (LTI) with first order transfer function and the controller is relay controller. In general the system is similar to the example system in section 2.2. The actuator in this figure includes "t" criteria in order to make decision. When the sensor outcome is over the limits, the sensor sends a message to the actuators. Actuators ask nodes 1 to t about their status. They reply and then the actuator makes decision. We suppose that the data exchange between the actuator and nodes 1 to t, happens with k hops.

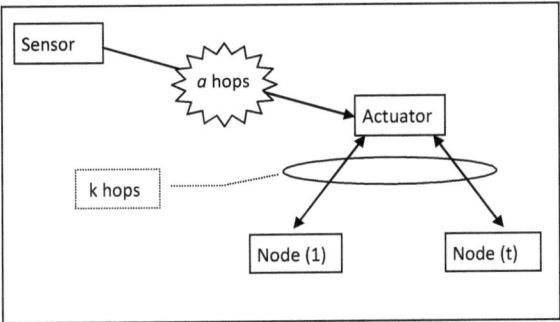

Figure 3.23: Autonomous communication path

The message number in time unit is calculated in Eq. 3.11. In this formula, the first component is the message number between the sensor and the actuator and the added component is the message number for data acquisition with k hops. It can be seen that the relation between message number and hop number is linear. By distancing the sensor and actuator from each other, the message umber increases linearly. In equation 3.11, f_{da} is the actuator frequency in discrete domain (after sampling). In section 2.2 it is shown how the actuator frequency can be calculated from the actuator frequency in the continuous domain and the actuator frequency drift. The computation formula is presented in Eq. 3.12. N_a in this formula is the optimal sample number for the model in Fig. 3.23 without criteria. As example suppose that T_n=3600, Y_{hc}=0.7, Y_{lc}=0.5 and a=5, then regarding to Table 2.3, N_a is equal to 23. By these values and k=0, the message number (f_{ma}) is equal to 0.005267 per time unit.

$$f_{ma} = 2 \times a \times f_{da} + 2 \times 2 \times f_{da} \times k \quad \text{Equation 3.11}$$

$$f_{da} = f_c \times (1 + p(N_a)) = \frac{f_s}{N_a} \times (1 + P(N_a)) \quad \text{Equation 3.12}$$

For the central network, the communication path in Fig. 3.24 is considered. The sensor sends a message at each sample period to the center. The center makes decision and sends message to the actuator at half of each actuator period. The message number in this structure is computed in Eq. 3.13. With C_{sc}=3, C_{ca}=8 and above mentioned system parameters, the message number is equal to 0.01737 (f_{mc}=0.01737) per time unit. As it is mentioned the criteria do not have any impacts on the message number, in other words it means the message number is constant versus hop number (criterion number).

$$f_{mc} = C_{sc} \times f_s + 2 \times f_{da} \times C_{ca} \quad \text{Equation 3.13}$$

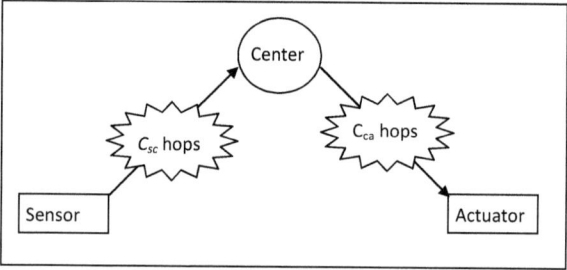

Figure 3.24: Central communication path

3.5.2) Comparison by message number

Now to compare the message number variations versus hop number in the central and autonomous networks, the Eq. 3.11 and Eq. 3.12 are sketched with above assumed parameters in Fig. 3.25. In this figure it can be seen when the hop number is zero, the message number in the autonomous network is less than the central one. By increasing the hops number, the message number linearly and gradually increases so that with 7 hops number, it becomes greater than message number in central network. Regarding to these figure necessity of having lower message number, the autonomous structure is

3. Simulation and Comparison

preferred to the central one when maximum number of hops for data acquisition is less than or equal to 6. As it is mentioned, the hops number is associated to criterion number exactly. These seven hops could be from just one criterion (one node) in Fig 3.11 or it could be distributed between more nodes, e.g. 1 hop for six nodes (six criteria) or two hops for one node and four hops for another one (two criteria). If the hops number in autonomous number becomes larger than six, then it can be said that central structure can function better than the autonomous one. It should be noted the advantages of distributed system in first chapter i.e. reliability is still valid and it is not related to this discussion. In this sense that in the central network if the center fails down the network will be down but in autonomous network by failing down a node the network can still function partly.

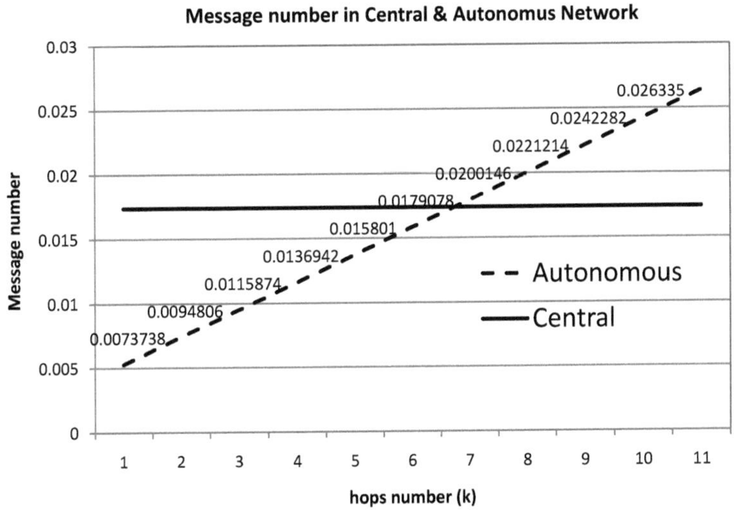

Message number in Central & Autonomus Network

hops number (k)	Autonomous	Central
1	0.0073738	
2	0.0094806	
3	0.0115874	
4	0.0136942	
5	0.015801	
6	0.0179078	
7	0.0200146	
8	0.0221214	
9	0.0242282	
10	0.026335	

Figure 3.25: Message number in exemplary Central and Autonomous network

3.5.3) Compromising based on the total message number

In above just one pair of associated sensor and actuator with their communication path is considered. Now to generalize the concepts and conclusions, firstly a network is decomposed to its associated sensor and actuator pairs. It is assumed that the number of these associated sensor and actuator pairs is equal to "m". Obviously the communication path model in the autonomous network will be like Fig. 3.23 and in the central network it will be the same model shown in Fig. 3.24. With this point of view in order to compute the total message number in these networks, the message numbers of all m paths are accumulated. In other words in each network the total message number will be equal to the summation of the message number of all paths. For the autonomous network, the total message number (F_{mta}) is calculated in Eq. 3.14. As it is taken into account each sensor-actuator pair has a designated optimal sample number (hidden in f_{da}^i) and its hop number (k_i). The hops number are taken differently because not necessarily all the paths have the same number of criteria and hops. For the central network, the total message number (F_{mtc}) is calculated in Eq. 3.15. In this equation F_s^i is the sensor optimal sample number and f_{da}^i is the actuator frequency in the discrete domain. C_{sc}^i and C_{ca}^i are orderly the corresponding hops number of the sensor-center and actuator-center wings.

$$F_{mta} = \sum_{i=1}^{m} f_{ma}^i = \sum_{i=1}^{m}(2 \times a_i \times f_{da}^i + 4 \times f_{da}^i \times k_i) \quad \text{Equation 3.14}$$

$$F_{mtc} = \sum_{i=1}^{m} f_{mc}^i = \sum_{i=1}^{m}(C_{sc}^i \times f_s^i + 2 \times f_{da}^i \times C_{ca}^i) \quad \text{Equation 3.15}$$

$$F_{mta}^{ref} = F_{mta}|_{k_i=0}$$
$$\frac{F_{mtc} - F_{mta}}{F_{mtc} - F_{mta}^{ref}} \quad \text{Equation 3.16}$$

On the contrary to the Eq. 3.11 relation between the total message number and hop number in Eq. 3.14 is not linear unless we assume that all the paths incorporate the same number of hops for data acquisition. Since application to application, the hop number could be different, this assumption is not realistic. For this reason in order to compare these two networks in general, not pair by pair, the total message number is taken as element for each network. For example in this way it can happen that the message number of one pair of sensor an actuator become greater in the autonomous network than the message number of the same pair in the central network but the total message number of the autonomous network become smaller the central network. In this case the autonomous structure can still be justified for the whole network.

In general if (F_{mtc}-F_{mta}) is positive, then the message number in the central network is more than the message number in the autonomous network, therefore it consumes

3. Simulation and Comparison

more energy and it can be affected more in the noisy environment. The maximum difference between both networks happens when the criterion number are zero in the autonomous network. According to this point the privilege of autonomous network to the central one can be expressed relatively. The formulation for such relative comparison is expressed in Eq. 3.16. For example if the outcome of Eq. 3.16 is 0.6, it means that the difference between autonomous and central network is 60 percent of the maximum value. When this percentage becomes smaller, it means the difference between the central and autonomous structures in terms of the factors which are dependent on message number becomes less too.

Chapter 4

Conclusion

In this chapter the brief conclusion of this work is outlined. In this work the autonomous system concept is discussed. Self-decision making of entities is recognized as the main property of this system. This property implementation is developed for the wireless sensor actuator network as the lowest layer of an automation system. The sensors make decisions as to when to send a message and actuators decide when to change their status by themselves. Achieving the self-decision making and enhancing reliability of the system necessitates that the entities have minimum dependency on each other or on a certain node. But in order to make decisions, it can occur that the entities require information from others. In this sense they become interdependent instead of dependent on resources of a central entity or entities in the higher level of hierarchical configuration. Dependency on a specific entity contradicts their self-decision making property because they can be controlled through this dependency as well. It is discussed that by making the resources autonomous entities, the competition can be forwarded to the other entities from resources to goals and conducts them to achieve a better performance. Interdependency, horizontal relations and transferring the decision level to the system peripherals make the system more robust in a risky environment while increasing the system reaction and self-adaptation speed against environmental changes.

A sequential coordinate routing algorithm is developed for the autonomous wireless sensor actuator network, whose main property is being target oriented. It means that, by this routing algorithm, any node is able to send a message to others directly. Other properties of this routing algorithm are easy computation, considering the minimum energy consumption, preserving the network scalability, no requirement of any embedded element and being reactive. The message is routed along the minimum length tree of the network and it does not face the Void problem either.

During this work we answered the question of how often the sensor should take samples from the system output in the autonomous and central network. This question was challenging because of the wireless communication character and network structure. A wireless node example is introduced by its hardware and software specification. By means of these nodes the autonomous wireless sensor actuator network is established. The main property of these nodes is being event-driven. It means they are normally in sleep mode whenever they receive a hardware or software interrupt, whereupon they wake up, provide the required service and then go back to sleep mode. The tiny operation system of this embedded system is real-time software.

Prowler is introduced and used as a simulator for the wireless sensor actuator network simulation. Its platform is changed to the introduced wireless node in this project and some other modifications are applied to this simulator. An energy model for the wireless transceiver is implemented in Prowler. This energy model consists of reception and multilevel transmission energy consumption. Simulations of autonomous and central networks for energy consumption evaluation led to two conclusions. Firstly, an autonomous network for performing a task consumes less energy than the central network. Secondly, the network energy consumption of performing a task is distributed over the nodes in the autonomous network more evenly in comparison to the central network. In other words some nodes in the central network are depleted much faster than others. This phenomenon causes the central network to be less sustainable. By generalizing this conclusion to the logistic system, the energy consumption can be replaced by fuel consumption in transportation.

In another simulation the robustness of these two networks is compared. It is concluded that the autonomous network is more robust than the central network in a noisy environment. In other words, the autonomous network resists and continues its functionality in a harsh environment whereas the central network fails. It is observed that the distribution of the noise over the autonomous network is more even. By resemblance and generalization, it is implied that the autonomous system can be more immune and achieve better service delivery when a high probability of transportation failure or risk exists in a logistic system. Following the optimal sample number method, it is shown that the autonomous network has the capability of scaling in contrast to the central network. Increasing the sample number in the autonomous network increases the process energy consumption, whereas in the central network the transmission energy consumption increases. Since the process energy consumption is much smaller than the transmission energy consumption, there is an option to choose a larger sample number in the autonomous network to achieve better control quality.

The validity of the above results is verified by looking at the diversity of the required information for making decisions. In the case that the nodes make decisions just based on received information, the above results are valid. The message number inside the network increases when the nodes require information from other nodes. This kind of decision making is called conditional decision making. It is shown that by raising the conditions number, at a specific point the message number in the autonomous network will be higher compared to the central network. The message number is related directly to the network energy consumption and its sensitivity to the noises in its environment. This specific point is computed for a single control feedback model and for the whole network.

References

[01King]- Preston King, Federalism and Federation, Taylor & Francis, 1982.

[02Wu]- Wu Jie, Distributed System Design, CRC-Press, 1998.

[03Dres]- Dressler Falko, Self-Organization in Sensors and Actor Networks, John Wiley & Sons Ltd, 2007.

[04Wang]- Wang Fei-Yue , Liu Derong, Networked Control Systems: Theory and Applications , Springer-Verlag London Limited, 2008.

[05Milan]- Gregory K. McMillan, Douglas M. Considine, Process/Industrial Instruments and Controls Handbook, Fifth Edition, McGraw-Hill Companies, 1999.

[06Zhan]- Peng Zhang, Industrial Control Technology, William Andrew Inc, 2008.

[07Lev]- William S. Leviene, The Control Handbook, Volume I, CRC press & IEEE Press, 1999.

[08CIS]- CISBE Guide H, Building Control Systems, The Chartered Institution of Building Services Engineers, 2000.

[09Duff]- Neil A. Duffie, Challenges in Design of Heterarchical Controls for Dynamics Logistic Systems, First International Conference on Dynamics Logistic, LDIC 2007, August 2007.

[10Dull]- Geir. E. Dullerud, G. Paganini Fernando, A Course in Robust Control Theory: A Convex Approach,Spriger-Verlag, 1999.

[11Lutz]- Lutz Rauchhaupt , System and Device Architecture of a Radio Based Fieldbus The RFieldbus System, 4th IEEE International Workshop on Factory Communication Systems, Västerás, Sweden, August 28-30, 2002

[12Tanb]- Tanenbaum Andrew s., Van steen Maarten , Distributed Systems Principles and Paradigms , First Edition, Prentice-Hall Inc , 2002.

[13will]- Andreas Willig, Kirsten Matheus, Adam Wolisz, Wireless Technology in Industrial Networks, Appeared in Proceedings of the IEEE, Vol. 93 (2005), No. 6 (June), pp. 1130-1151.

[14King]- Preston King, Federalism and Federation, Taylor & Francis, 1982.

[15Elaz]- Daniel J.Elazar, Exploring Federalism, University of Alabama Press, 1987.

[16cont]- R. Jedermann, C.Behrens, R.Laur,W. Lang, Intelligent containers and sensor networks, Approaches to apply autonomous cooperation on systems with limited resources. In: Hülsmann, M.; Windt, K. (eds.): Understanding Autonomous Cooperation & Control in Logistics - The Impact on Management, Information and Communication and Material Flow. Springer, Berlin, 2007, Pages: 365-392.

[17gpsr]- Brad Karp, H. T. Kung, GPSR: Greedy Perimeter Stateless Routing for Wireless Networks, International Conference on Mobile Computing and Networking Proceedings of the 6th annual international conference on Mobile computing and networking, Boston, Massachusetts, United States, Year of Publication: 2000. Pages: 243 – 254.

[18lcr]- Qing Cao, Tarek Abdelzaher, A Scalable Logical Coordinates Framework for Routing in Wireless Sensor Networks, Department of Computer Science, University of Virginia, Charlottesville, VA 22904, Proceed ings of the 25th IEEE International Real-Time Systems Symposium, 2004, USA, Date: 5-8 Dec. 2004, Pages: 349 – 358.

[19gear]- Yan Yu, Ramesh Govindan, Deborah Estrin, Geographical and Energy Aware Routing: a recursive data dissemination protocol for wireless sensor networks, In Proceedings. ACM Mobicom, Boston MA, 2001.

[20leac]- Heinzelman, W.R., Chandrakasan, A., Balakrishnan, H, Energy-efficient communication protocol for wireless microsensor networks, System Sciences, 2000, Proceedings of the 33rd Annual Hawaii , Date: 4- 7 Jan. 2000, Pages: 10-17.

[21zrp]- Nicklas Beijar, Zone Routing Protocol (ZRP), Networking Laboratory, Helsinki University of Technology,Finland.

[22aodv]- Perkins, C.E., Royer, E.M., Ad-hoc on-demand distance vector routing. Mobile Computing Systems and Applications, 1999. Proceedings. WMCSA '99. Second IEEE Workshop on Date: 25-26 Feb 1999, Pages: 90 – 100.

[23grad]- GRAdient Broadcast: A Robust, Long-lived Large Sensor Network. http://irl.cs.ucla.edu/papers/grab-tech-report.ps.

[24tanb]- Andrew S.Tanenbaum, Computer Networks, 4 edition, Prentice Hall PTR, August 9, 2002.

[25fusi] - Heinzelman, W.R., Chandrakasan, A., Balakrishnan, H, "Energy-efficient communication protocol for wireless microsensor networks", System Sciences, 2000, Proceedings of the 33rd Annual Hawaii , Date: 4- 7 Jan. 2000, Pages: 10.

[26gary]- Gary Chartrand, Ortrud R. Ollermann, Applied and algorithmic Graph Theory McGraw-Hill, 1993.

[27foul]- L.R. Foulds, Graph Theory Applications, Springer-Verlag, 1992.

[28jung]- Dieter Jung nickel, Graphs, networks and Algorithms, Third Edition, Springer-Verlag, 2008.

[29mhr]- Heinzelman, W.R.; Chandrakasan, A. Balakrishnan, H: Energy-efficient communication protocol for wireless microsensor networks, Proceedings of the 33rd Annual Hawaii International on System Sciences, Pages: 10 vol.2. IEEE Press (2000).

[30154]- LAN MAN Standards committee of the IEEE Computer Society, IEEE, Wireless Medium Access Control (MAC) and Physical Layer (PHY) Specifications for Low-Rate Wireless Personal Area Networks (LR-WPANs), NewYork, NY, USA.IEEE Std 802.15.4™-2003.

[31msp]- R. G. Gallager, P. A. Humblet, and P. M. Spira, A Distributed Algorithm for Minimum-Wight Spanning Trees, ACM Transactions on Programming Languages and Systems, Vol. 5, No. 1, January 1983, Pages 66- 77.

[32anto]- Antonio Caruso, Stefano Chessa, and Swades De, Relation Between Gradients and Geographic Distances in Dense Sensor Networks with Greedy Message Forwarding, Fourth International

Confernce on Systems and Networks Communications, 2009, ICSNC '09, 20-25 Sept. 2009 Pages: 236-241.

[33fang]- Q. Fang, J. Gao, L. J. Guibas, V. de Silv, and L. Zhang. Glider: Gradient landmark-based distributed routing for sensor networks, in Proc, IEEE INFOCOM, March 2005.

[34sca1]- Amir M. Jafari, Adam Sklorz, and Walter Lang, Target-Oriented Routing Algorithm Based on Sequential Coordinates for Autonomous Wireless Sensor Network, Journal of Networks, Academy Publisher, VOL. 4, NO. 6, August 2009, Pages: 421-427.

[35sca2]- Amir Jafari, Adam Sklorz, and Walter Lang," SCAR: Sequential Coordinate Routing Algorithm for Autonomous Wireless Sensor Network", the 2nd International Conference on New Technologies, Mobility and Security (NTMS 2008), Tangier, Morocco, 5-7, November 2008, Pages: 427-432.

[36Gay]- David Gay, et al, nesC 1.1 Language Reference Manual, http://nescc.sourceforge.net/papers/nesc- ref.pdf, May 2003.

[37tiny]- TinyOS Tutorial, www.tinyos.net, 2003.

[38sky]- Moteiv Corporation, tmote-sky-datasheet-102, www.moteiv.com, 2006.

[39msp]- Texas Instruments Incorporated, MSP430f1611 Data Sheet, www.ti.com, 2005.

[40msp]- Texas Instruments Incorporated, User's Guide, MSP430x1xx Family, www.ti.com, 2006.

[41open]- Alan V. Oppenheim, Ronald W.Schafer, John R. Buck, Discrete-Time Signal Processing, 2nd Edition, Prentice Hall, January 10, 1999.

[42toss]- Philip Levis, Nelson Lee, Matt Welsh, and David Culler TOSSIM: Accurate and Scalable Simulation of Entire TinyOS Applications. In the Proceedings of First ACM Conference on Embedded Networked Sensor systems (SenSys 2003).

[43sidh]- Thomas W. Carley, "Sidh: A Wireless Sensor Network Simulator", Department of Electrical & Computer Engineering University of Maryland at College Park. ISR technical Reports,2004.

[44prow]- Gyula Simon, Péter Völgyesi, Miklós Maróti, Ákos Lédeczi, "Simulation-based optimization of communication protocols for large-scale wireless sensor networks", Submitted to 2003 IEEE Aerospace Conference, Big Sky, MT, March 8-15, 2003.

[45cc24]- Chipcon AS, CC2420 Data Sheet, www.chipcon.com , 2006.

[46ogat]- Katsuhiko Ogata, System Dynamics, Forth edition, Prentice-Hall Inc, 2004.

[47Böse]- Felix Böse, Katja Windt, Catalogue of Criteria for Autonomous Control in Logistics, In: Hülsmann, M.; Windt, K. (eds.): Understanding Autonomous Cooperation & Control in Logistics - The Impact on Management, Information and Communication and Material Flow. Springer, Berlin, 2007, Pages: 57-72.

[48eng]- Amir Jafari, Adam Sklorz, and Walter Lang, "Energy Consumption Comparison between Autonomous and Central Wireless Sensor Network", International Conference on Mobile Communications and Pervasive Computing (MCPC 2009), the Second SIWN Congress (SIWN 2009), Leipzig, Germany, 23-25 March 2009.

[49opt]- Amir Jafari, and Walter Lang, "Optimal Sample Number for Central and Autonomous Wireless Sensor Network in Process Automation Applications", World Congress on Engineering 2009

(WCE 2009), The International Conference of Wireless Networks (ICWN 2009), London, U.K., 1-3 July 2009, Pages: 869-874.

[50robs]- Amir Jafari, Dirk Hentschel and Walter Lang, "Robustness in Autonomous and Central Wireless Sensor Network: The Orchard Example", Fourth International Conference on Systems and Networks communications 2009 (ICSNC 2009), IEEE computer Society, Porto, Portugal, 20-25 September 2009, Pages: 242-247.

[51eng]- Amir Jafari, Adam Sklorz and Walter Lang, "Energy Consumption Comparison Between Autonomous and Central Wireless Sensor Network", In: "Communications of SIWN", ISSN 1757-4439 (Print), ISSN 1757-4447 (CD-ROM) Vol. 6, April 2009, Pages: 166-170.

[52opt]- Amir Jafari and Walter Lang, "Optimal Sample Rate for Wireless Sensor Actuator Network", In: "IAENG International Journal of Computer Science", ISSN: 1819-9224 (online version); 1819-656X (print version), Volume: 36, Issue: 4, November 2009, Pages: 387-393.

Publications List

Books

[1]- Amir Jafari, Walter Lang, "Optimal Sample Number For Autonomous And Central Wireless Sensor Actuator Network" In: "Electronic Engineering and Computing Technology", by: Ao, Sio-long; Gelman, Len; (eds.), Springer,1st Edition,2010, ISBN: 978-90-481-8775-1.

Journals

[2]- Amir M Jafari, Adam Sklorz, Walter Lang, Energy Consumption Comparison between Autonomous and Central Wireless Sensor Network. In: "Communications of SIWN", ISSN 1757-4439 (Print), ISSN 1757-4447 (CD-ROM) Vol. 6, April 2009, Pages: 166-170.

[3]- Amir Jafari, Adam Sklorz, Walter Lang, Target-Oriented Routing Algorithm Based on Sequential Coordinates for Autonomous Wireless Sensor Network, In: "Journal of Networks", ISSN: 1796-2056, Volume: 4, Issue: 6, August 2009, Pages: 421-427.

[4]- Amir M Jafari, Walter Lang, Optimal Sample Rate for Wireless Sensor Actuator Network, In: "IAENG International Journal of Computer Science", ISSN: 1819-9224 (online version); 1819-656X (print version), Volume: 36, Issue: 4, November 2009, Pages: 387-393.

Proceedings

[5]- Amir Jafari, Adam Sklorz, Lang W. :"SCAR: Sequential Coordinate Routing Algorithm for Autonomous Wireless Sensor Network", NTMS Int. Conf. on New Technologies, Mobility and Security (NTMS08),IEEE ESRGroups France, Tangier, Morocco, November 5-7, 2008, pages: 1-6.

[6]- Amir Jafari, Adam Sklorz, Walter Lang, "Energy Consumption Comparison between Autonomous and Central Wireless Sensor Network", International Conference on Mobile Communications and Pervasive Computing (MCPC 2009) The Second SIWN Congress (SIWN 2009), Leipzig, Germany, 23-25 March 2009.

[7]- Amir Jafari, Walter Lang, "Optimal Sample Number for Central and Autonomous Wireless Sensor Network in Process Automation Applications", World Congress on Engineering 2009 (WCE 2009), The International Conference of Wireless Networks (ICWN 2009, London, U.K., 1-3 July 2009, Pages: 869-874.

[8]- Amir Jafari, Dirk Hentschel, Walter Lang, "Robustness in Autonomous and Central Wireless Sensor Network: The Orchard Example", Fourth International Conference on Systems and Networks Communications 2009 (ICSNC 2009), IEEE computer Society, Porto, Portugal, 20-25 September 2009, Pages: 242-247.

Die VDM Verlagsservicegesellschaft sucht für wissenschaftliche Verlage abgeschlossene und herausragende

Dissertationen, Habilitationen, Diplomarbeiten, Master Theses, Magisterarbeiten usw.

für die kostenlose Publikation als Fachbuch.

Sie verfügen über eine Arbeit, die hohen inhaltlichen und formalen Ansprüchen genügt, und haben Interesse an einer honorarvergüteten Publikation?

Dann senden Sie bitte erste Informationen über sich und Ihre Arbeit per Email an *info@vdm-vsg.de*.

Sie erhalten kurzfristig unser Feedback!

VDM Verlagsservicegesellschaft mbH
Dudweiler Landstr. 99   Telefon +49 681 3720 174
D - 66123 Saarbrücken   Fax   +49 681 3720 1749
www.vdm-vsg.de

Die VDM Verlagsservicegesellschaft mbH vertritt

Printed by Books on Demand GmbH, Norderstedt / Germany